TEV

The Reservoir Engineering Aspects of Waterflooding

Forrest F. Craig, Jr.

Manager, Petroleum Engineering

Amoco International Oil Company

Third Printing
November 1980

Henry L. Doherty Memorial Fund of AIME
Society of Petroleum Engineers of AIME

New York   1971   Dallas

DEDICATION

To

Betty, Mom, Forrest III, and Elizabeth Anne

and

To Those Many Friends From Whom I've Learned

© Copyright 1971 by the American Institute of Mining, Metallurgical, and Petroleum Engineers, Inc. Printed in the United States of America by Millet the Printer, Dallas, Tex. All rights reserved. This book, or parts thereof, cannot be reproduced without written consent of the publisher.

ISBN 0-89520-202-6

Contents

1. **Introduction** — 9
   - 1.1 General History and Development of Waterflooding — 9
   - 1.2 Relation Between Reservoir Engineering and Production Engineering Aspects of Waterflooding — 9
   - 1.3 Objectives of the Monograph — 10
   - 1.4 Organization of the Monograph — 10

2. **Basic Water-Oil Flow Properties of Reservoir Rock** — 12
   - 2.1 Rock Wettability — 12
   - 2.2 Fluid Distribution — 15
   - 2.3 Capillary Pressure — 17
   - 2.4 Relative Permeability — 19
   - 2.5 Native-State and Restored-State Cores — 21
   - 2.6 Methods of Measuring Water-Oil Flow Properties — 22
   - 2.7 Three-Phase Relative Permeabilities — 24
   - 2.8 Connate Water Saturation — 25

3. **Efficiency of Oil Displacement by Water** — 29
   - 3.1 Frontal Advance Theory — 29
   - 3.2 Water Tonguing — 34
   - 3.3 Viscous Fingering — 34
   - 3.4 Mobility of the Connate Water — 34
   - 3.5 Practical Application of the Frontal Advance Theory — 35
   - 3.6 Influence of Rock Wettability on Oil Production Performance — 38
   - 3.7 Influence of Oil and Water Viscosities — 38
   - 3.8 Influence of Formation Dip and Rate — 39
   - 3.9 Influence of Initial Gas Saturation — 39

4. **Mobility Ratio Concept** — 45
   - 4.1 Development of the Mobility Ratio Concept — 45
   - 4.2 Definition of Mobility Ratio — 45
   - 4.3 Ranges of Mobility Ratios During Waterflooding — 46

5. **Areal Sweep Efficiency** — 48
   - 5.1 Definition — 48
   - 5.2 Measurement — 48
   - 5.3 Areal Sweep Efficiency at Water Breakthrough — 50
   - 5.4 Areal Sweep Increase After Water Breakthrough — 53
   - 5.5 Injectivities in Various Waterflood Patterns — 53
   - 5.6 Areal Sweep Prediction Methods — 54
   - 5.7 Other Factors Affecting Areal Sweep — 56
   - 5.8 Factors Affecting Selection of Waterflood Pattern — 58

6. **Reservoir Heterogeneity** — 62
   - 6.1 Types of Reservoir Heterogeneities — 62
   - 6.2 Quantitative Descriptions of Permeability Stratification — 63

7. **Vertical and Volumetric Sweep Efficiencies** — 69
   - 7.1 Definition — 69
   - 7.2 Influence of Mobility Ratio — 69
   - 7.3 Influence of Gravity Forces — 71
   - 7.4 Influence of Capillary Forces — 72
   - 7.5 Crossflow Between Layers — 73
   - 7.6 Effect of Rate on Volumetric Sweep Calculations — 75
   - 7.7 Effect of Layer Selection on Volumetric Sweep Calculations — 75

8. **Methods of Predicting Waterflood Performance** — 78
   - 8.1 The Perfect Prediction Method — 78
   - 8.2 Prediction Methods Primarily Concerned with Reservoir Heterogeneity — 78

8.3 Prediction Methods Primarily Concerned with Areal Sweep ... 79
8.4 Prediction Methods Dealing Primarily with Displacement Mechanism ... 81
8.5 Prediction Methods Involving Mathematical Models ... 82
8.6 Empirical Waterflood Prediction Methods ... 83
8.7 Comparisons of Performance Prediction Methods ... 84
8.8 Comparisons of Actual and Predicted Performance ... 87
8.9 Recommended Waterflood Prediction Methods ... 90
8.10 Practical Use of Waterflood Prediction Methods ... 93
8.11 Factors Affecting Waterflood Oil Recovery Performance ... 94

9. Pilot Waterflooding ... 97
9.1 Advantages and Limitations of Pilot Floods ... 97
9.2 Information Obtainable from Pilot Waterfloods ... 97
9.3 Pilot Flood Design ... 99

10. Conclusion ... 101
10.1 The State of the Art ... 101
10.2 Current Problems and Areas for Further Study ... 101

Appendix A: Derivation of Fractional Flow Equation ... 103

Appendix B: Derivation of the Frontal Advance Equation ... 105

Appendix C: Alternative Derivation of Welge's Equation ... 107

Appendix D: Additional Published Design Charts and Correlations ... 108

Appendix E: Example Calculations ... 112
E.1 Calculation of Fractional Flow Curve and Displacement Performance ... 112
E.2 Calculation of Mobility Ratio ... 114
E.3 Calculation of Ultimate Waterflood Recovery ... 114
E.4 Composite WOR-Recovery Performance ... 114
E.5 Composite Injection and Producing Rates, WOR, and Recovery vs Time ... 115
E.6 Individual Well Performance ... 123

Nomenclature ... 125

Bibliography ... 127

Subject-Author Index ... 135

SPE Monograph Series

The Monograph Series of the Society of Petroleum Engineers of AIME, one of SPE's more recent publication programs, was established in 1965 by action of the SPE Board of Directors. This Series is intended to provide members with an authoritative, up-to-date treatment of the fundamental principles and state of the art in selected fields of technology. The work is directed by the Society's Monograph Committee, one of 36 national committees. Below is a roster of committee members.

1965
J. M. C. Gaffron, Chairman, Freeport Oil Co., New Orleans
Buck Joe Miller, Petroleum Consultant, Dallas
Henry J. Ramey, Jr., Stanford U., Stanford, Calif.
J. G. Richardson, Humble Oil & Refining Co., Houston
Shofner Smith, Phillips Petroleum Co., Bartlesville, Okla.
Bruce Vernor, Atlantic Richfield Co., New York City
M. R. J. Wyllie, Gulf Research & Development Co., Pittsburgh

1966
J. M. C. Gaffron, Chairman, Freeport Oil Co., New Orleans
Buck Joe Miller, Petroleum Consultant, Dallas
Don G. Russell, Shell Oil Co., New York City
M. R. J. Wyllie, Gulf Research & Development Co., Pittsburgh
Robert S. Cooke, Union Oil Co. of California, Midland, Tex.
Henry J. Ramey, Jr., Stanford U., Stanford, Calif.
Charles Blomberg, Chevron Oil Co., New Orleans
F. F. Craig, Jr., Pan American Petroleum Corp., Tulsa

1967
Henry J. Ramey, Jr., Chairman, Stanford U., Stanford, Calif.
M. R. J. Wyllie, Gulf Research & Development Co., Pittsburgh
Robert S. Cooke, Union Oil Co. of California, Midland, Tex.
Don G. Russell, Shell Oil Co., New York City
J. R. Elenbaas, Michigan Consolidated Gas Co., Detroit
B. F. Burke, Humble Oil & Refining Co., Long Beach, Calif.
B. L. Sledge, Humble Oil & Refining Co., New Orleans
C. Max Stout, Dowell, Tulsa

1968
Don G. Russell, Chairman, Shell Oil Co., New York City
Robert S. Cooke, Union Oil Co. of California, Midland, Tex.
J. R. Elenbaas, Michigan Consolidated Gas Co., Detroit
B. F. Burke, Humble Oil & Refining Co., Long Beach, Calif.
William E. Brigham, Continental Oil Co., Ponca City, Okla.
J. G. Richardson, Humble Oil & Refining Co., Houston
Lowell Murphy, Eppler, Guerin & Turner, Dallas
Roland F. Krueger, Union Oil Co. of California, Brea, Calif.

1969
B. F. Burke, Chairman, Humble Oil & Refining Co., Long Beach, Calif.
J. R. Elenbaas, Michigan Consolidated Gas Co., Detroit
William E. Brigham, Continental Oil Co., Ponca City, Okla.
J. G. Richardson, Humble Oil & Refining Co., Houston
James H. Henderson, Gulf Research & Development Co., Houston
Oren C. Baptist, U. S. Bureau of Mines, San Francisco
Richard L. Perrine, UCLA, Los Angeles
Robert A. Wattenbarger, Scientific Software Corp., Englewood, Colo.

1970
J. G. Richardson, Chairman, Humble Oil & Refining Co., Houston
William E. Brigham, Continental Oil Co., Ponca City, Okla.
Oren C. Baptist, U. S. Bureau of Mines, San Francisco
James H. Henderson, Gulf Research & Development Co., Houston
W. T. Bell, Schlumberger Well Services, Houston
Roland F. Krueger, Union Oil Co. of California, Brea, Calif.
George H. Bruce, Esso Production Research Co., Houston
Lincoln F. Elkins, Sohio Petroleum Co., Oklahoma City

1971
Oren C. Baptist, Chairman, U. S. Bureau of Mines, San Francisco
James H. Henderson, Gulf Oil Corp., Houston
George H. Bruce, Esso Production Research Co., Houston
Roland F. Krueger, Union Oil Co. of California, Brea, Calif.
W. B. Gogarty, Marathon Oil Co., Littleton, Colo.
Necmettin Mungan, Petroleum Recovery Research Institute, Calgary
Michael Prats, Shell Development Co., Houston
A. B. Waters, Halliburton Services, Duncan, Okla.

Forrest F. Craig Jr.
(1926-1978)
In Memoriam

Forrest F. (Woody) Craig Jr., 1977 SPE president, was known internationally as author, lecturer, and energetic advocate of SPE programs. He received the Society's John Franklin Carll Award in 1977 for his contributions to the petroleum engineering profession, especially in waterflood and reservoir performance prediction techniques, and his active participation in all levels of Society affairs.

Craig received BS, MS, and PhD degrees in chemical engineering from the U. of Pittsburgh. He joined Stanolind Oil and Gas Co. (an Amoco predecessor) in 1951 as research engineer with the Production Research Div. in Tulsa. In 1960 he began supervising research in reservoir performance prediction, thermal recovery techniques, and field testing of unconventional oil recovery methods. During the 1960's, Craig received eight patents on miscible and thermal recovery methods and testified at 12 U.S. regulatory agency hearings. In 1971 he moved to Chicago to become chief petroleum engineer with Amoco Intl. Oil Co. Amoco named him worldwide petroleum engineering manager in 1975.

He contributed 18 technical works to Society literature since 1952, and was involved in Society and Mid-Continent Section activities. Craig chaired the 1964 Annual Meeting Program, 1966 Transactions Editorial, and 1969 Continuing Education committees. He was 1967-68 chairman of the Mid-Continent Section in Tulsa.

The first printing of this Monograph in 1971 opened a new decade of Society involvement. Craig served as program chairman of the 1972 Symposium on Improved Oil Recovery Methods and chairman of the 1972 Distinguished Service Award Committee, the 1973 *ad hoc* Committee To Study Overseas Policy, and the 1975 *ad hoc* Committee on Study of Field Case History and Transactions Papers. He was 1973-74 Chicago Petroleum Section chairman.

Craig served continuously on the SPE board since becoming a regional director in 1973. He later was a director at large. In Oct. 1976, he became the first Society president to be installed at the Annual Meeting. Among his goals as president were improving communication among local sections through the officer visitation program, and balanced delivery of Society services to all SPE sections worldwide.

He remained active in SPE-AIME affairs until his death, serving on the 1978 SPE and AIME boards, as Nominating Committee chairman, and as a member of the Executive Committee. Craig died March 8, 1978, in Tulsa at age 52.

Preface

As any publication reflects the author's understanding of the subject, this Monograph reflects my knowledge of waterflooding. This knowledge was acquired over the years by study, by reading, by experience, but most important perhaps by discussion with my co-workers. It seems appropriate at this point, when the Monograph is in its final stages, to express a special thanks to those from whom I've learned. Included in this long list are Lloyd Elkins, Ted Geffen, Bill Tracy, Bill Owens, Jerry Neil, Don Henley, Bill Nation, Howard Hall, Dave Parrish, Dick Morse, the late Charlie Stewart, and others with whom I've worked, as well as those from many other companies.

The Society's Monograph Committee played a highly important part in the content of this Monograph. Their thorough review and constructive suggestions were a valuable help in achieving the balance and coverage of the subject.

I am very grateful to David Riley and Dan Adamson of the Society of Petroleum Engineers in Dallas for their confident encouragement during the many months of preparation of this publication and to Ann Gibson for her smooth, efficient handling of the production. My special thanks go to Sally Wiley for her perceptive job in editing the manuscript, for bringing into focus the sometimes fuzzy writing of an engineer.

Lastly, I owe my family an expression of my sincere gratitude for their patience, understanding, and encouragement during those long, lonely hours I spent planning and writing this Monograph.

F. F. CRAIG, JR.

December, 1970

Chapter 1

Introduction

1.1 General History and Development of Waterflooding

Waterflooding is dominant among fluid injection methods and is without question responsible for the current high level of producing rate and reserves within the U. S. and Canada. Its popularity is accounted for by (1) the general availability of water, (2) the relative ease with which water is injected, owing to the hydraulic head it possesses in the injection well, (3) the ability with which water spreads through an oil-bearing formation and (4) water's efficiency in displacing oil.

It is generally acknowledged that the first waterflood occurred as a result of accidental water injection in the Pithole City area of Pennsylvania in 1865.[1] In 1880 John F. Carll[2] concluded that water, finding its way into a wellbore from shallow sands, would move through oil sands and be beneficial in increasing oil recovery. Many of the early waterfloods occurred accidentally by leaks from shallow water sands or by surface water accumulations entering drilled holes. At that time it was felt that the main function of water injection was to maintain reservoir pressure, allowing wells to have a longer productive life than by pressure depletion.

In the earliest method of waterflooding, water was injected first at a single well; as the water-invaded zone increased and adjacent wells were watered out, these were used as injectors to extend the area of water invasion. This was known as "circle flooding".[1] As a modification of this technique, Forest Oil Corp.[3] converted a series of wells to water injection at one time, forming a line drive. The first five-spot pattern flood was attempted in the southern part of the Bradford field in 1924.[3]

The Bradford field had a large productive area, little gas in solution and no water encroachment. These factors contributed to the rapid growth of waterflooding there. Operators were slow to apply their waterflood experiences to areas outside Pennsylvania. In 1931 a waterflood was initiated in the shallow Bartlesville sand in Nowata County, Okla., and within a few years many of the Bartlesville sand reservoirs were under flood. The first waterflood in Texas was started in the Fry pool of Brown County in 1936. Within 10 years waterflooding was in operation in most of the oil producing areas. However, it was not until the early 1950's that the general applicability of waterflooding was recognized.

1.2 Relation Between Reservoir Engineering and Production Engineering Aspects of Waterflooding

In many companies that are planning and operating waterfloods, two functional organizations are involved. One deals with reservoir engineering and the other with production engineering.

Typically the reservoir engineers are responsible for all facets of work leading to the prediction of oil recovery performance. They are charged with accumulating all the necessary basic data for the reservoir in question. Frequently working with geologists or evaluation engineers, they prepare structure, isopach, and isobar maps of the reservoir. They assemble the available core analysis and fluid property data as well as the results of special well log or pressure transient studies. The reservoir engineers may request special laboratory tests to measure, for example, relative permeability or capillary pressure characteristics. Utilizing these data, they investigate various flooding patterns, select locations for water injection wells, estimate their injectivity, and recommend any additional development wells. Their end product is a detailed prediction of the oil recovery performance by waterflooding. This then becomes the basis for an economic projection of the profitability of the waterflood.

The production engineers, frequently in so-called "operations" groups, have been working along with the reservoir engineers, contributing their expertise in operational matters. This frequently involves selection and testing of water supply sources, design and sizing of water treating equipment, specification of metering and testing facilities, investigation of corrosion or scaling tendencies, and review of existing wells to determine any needed remedial work.

After a waterflood project has been approved, the engineers in both the reservoir engineering and operations sections share the responsibility. Those in the former are responsible for a continuous review of reservoir performance, updating and modifying expected performance. Those in the latter are responsible for the operation of surface injection and production facilities. Well workovers are jointly planned.

The roles depicted above for the reservoir engineer and his operational counterpart are oversimplified. Each has his own area of responsibility, but their cooperative effort is necessary for a successful waterflooding operation.

1.3 Objectives of the Monograph

This Monograph is devoted to the reservoir engineering aspects of waterflooding, which in simple terms means a quantitative understanding of how water displaces oil from reservoir rock. This includes the prediction of the water injection rates, oil producing rates, producing water-oil ratios, and cumulative oil recovery at various times in the future.

Reservoir engineering is one of the few applied sciences that deal with a system that cannot in its entirety be seen, weighed, measured, or tested. Even in fields where every drilled well is cored, less than one-millionth of the reservoir rock is ever sampled and seen by man. Fluid samples on which detailed laboratory measurements are made are similarly limited.

However limited the samples of reservoir rocks, diagnostic and interpretive techniques have been developed for gaining further information about the reservoir. Well logging techniques are comparatively inexpensive and frequently can be used in wells drilled years ago to obtain in-place measures of porosity, permeability, saturation, and lithological changes. These logging techniques measure the electrical, acoustical, or radioactive response of the formation within a radius several times the wellbore radius. Well pressure transient tests and their interpretation[4] have proved a versatile tool in measuring gross characteristics of formations. A yet newer technique is the application of geological engineering, in which information obtained from cores, logs, and bit cuttings is combined statistically to yield a detailed description of the reservoir permeability variation.[5] This approach is highly promising when applied to sandstones. Sands are generally well ordered; and relatively large areal segments of these formations are deposited at one time, with porosity existing at deposition. Carbonate formations in which porosity is developed after deposition by dolomitization and ground water movement contain more random variations in rock properties. In these formations the inference of interwell properties from measurements at wells is generally less reliable.

We are better able to simulate reservoir performance, it has been said, than to describe the reservoir itself. Currently our knowledge of the mechanism of oil displacement and our methods of calculating oil recovery performance, though not complete in every detail, exceed our abilities to describe the spatial variations in rock properties and fluid saturations within the reservoir. Nevertheless, though imperfect at present, reservoir engineering is indeed a science.

One objective of this Monograph is to provide a comprehensive presentation and critique of the available information dealing with the reservoir engineering aspects of waterflooding. This Monograph is aimed at the practicing engineer, allowing him to re-educate himself on the subject, and at the same time serving as an objective and authoritative reference book. The detailed mathematical derivations and example calculations are presented in the Appendices, so that the Monograph itself can emphasize the use of these equations to obtain a quantitative understanding of waterflooding.

Where different techniques involving a particular measurement or calculation are involved, a preferred method will be presented, together with an explanation for its preference. Alternative methods also will be discussed, sometimes in detail. In particular in the area of waterflood predictive methods, there are a multitude of variations. However, each main type will be discussed in detail with significant variations referenced. In this manner the practicing engineer will find a recommended technique, and the more studious engineer will be guided to the alternative procedures.

1.4 Organization of the Monograph

In this Monograph the presentation will lead from the fundamentals of waterflooding to the engineering application of this oil recovery method. The next chapter presents a discussion of the fundamental water-oil flow properties of reservoir rock and the factors that affect these properties. The problems involved in adequate reservoir sampling are reviewed. The third chapter describes the calculation of oil displacement efficiency from basic flow properties. Subsequent chapters discuss the mobility ratio concept and show its effect on the areal sweep efficiency of different injection-production well arrangements. Following this, the several types of reservoir heterogeneities are described in general terms, with emphasis on methods of characterizing vertical permeability variations. With this as a prelude, the effect of various factors on vertical sweep efficiency is then described. These discussions of displacement efficiency as well as areal and vertical sweep efficiencies logically lead to a description of some published methods for predicting waterflood oil recovery performance. Special emphasis is placed on the data required for these methods. Their capabilities and limitations are discussed, and as a means of evaluating them they are compared with a yet-to-be-achieved "perfect" prediction method. The ninth chapter deals with the interpretation of pilot performance. The final chapter presents an analysis of the current level of waterflooding technology, together with a definition of areas for further study.

Detailed mathematical derivations are presented in the Appendices, which also contain correlations and design charts useful in the prediction of waterflood performance. To illustrate the use of equations and correlations, the final Appendix presents a series of example problems. I sincerely hope that this manner of presentation will yield a readable, understandable, and practical guide to the reservoir engineering aspects of waterflooding.

References

1. *History of Petroleum Engineering*, API, Dallas, Tex. (1961).
2. Carll, John F.: *The Geology of the Oil Regions of Warren, Venango, Clarion, and Butler Counties, Pennsylvania*, Second Geological Survey of Pennsylvania (1880) **III,** 1875-1879.
3. Fettke, C. R.: "Bradford Oil Field, Pennsylvania and New York", Pennsylvania Geological Survey, 4th Series (1938) **M-21**.
4. Matthews, C. S. and Russell, D.G.: *Pressure Buildup and Flow Tests in Wells*, Monograph Series, Society of Petroleum Engineers, Dallas, Tex. (1967) **1**.
5. Alpay, O.: "A Study of Physical and Textural Heterogeneity in Sedimentary Rocks", PhD dissertation, Purdue U., Lafayette, Ind. (June, 1963).

Chapter 2

Basic Water-Oil Flow Properties of Reservoir Rock

A prerequisite for understanding waterflood performance is a knowledge of the basic properties of reservoir rock. These consist of two main types: (1) properties of the rock skeleton alone, such as porosity, permeability, pore size distribution, and surface area, and (2) combined rock-fluid properties such as capillary pressure (static) characteristics, and relative permeability (flow) characteristics.

At this point some basic definitions—widely accepted and pertinent to this Monograph—are in order.

Absolute Permeability—permeability of rock saturated completely with one fluid.

Effective Permeability—permeability of rock to one fluid when the rock is only partially saturated with that fluid.

Relative Permeability—ratio of effective permeability to some base value.

Porosity—portion of rock bulk volume composed of interconnected pores.

Little more will be said here about the basic properties of the rock skeleton. The definitions of permeability and porosity, as well as accepted techniques for their measurement, are treated effectively by Muskat,[1] Pirson,[2] and Calhoun.[3] Other rock properties such as pore size distribution and surface area also have been discussed in the literature. Correlations between pore size distribution and both permeability and relative permeability have been made.[4-6] However, although these correlations provide some understanding of microscopic flow behavior, they are subject to severe limitations and thus are seldom used in applied reservoir engineering.

The multiphase static and flow properties of reservoir rock listed above depend upon the microscopic distribution of these phases within the rock pores. This distribution is controlled by the rock wettability—that is, the degree of preference of the rock surface for the various fluid phases. Therefore, the related subjects of rock wettability and fluid distribution are the first to be discussed in this chapter.

2.1 Rock Wettability

Wettability is a widely used term. It can be defined as the tendency of one fluid to spread on or adhere to a solid surface in the presence of other immiscible fluids. Applying this term to reservoir engineering, the solid surface is the reservoir rock, sandstone, limestone, or dolomite (and often detrital or cementing material, or both). The fluids that exist in the rock pore spaces during waterflooding are oil, water, and gas. However, since the conditions under which gas, in preference to liquid,[7] will wet the rock surface, are beyond the range encountered in waterflooding reservoir rocks, only oil and water can be considered the possible wetting phases. The effect of a rock's wettability preference for either water or oil on the flow properties during waterflooding will be discussed in Section 2.4.

The evaluation of reservoir wettability has been discussed in detail.[8] Let us consider in an idealized way a typical water-oil-solid system as shown in Fig. 2.1. The surface energies in such a system are related by the Young-Dupre[9] equation as follows:

$$\sigma_{os} - \sigma_{ws} = \sigma_{ow} \cos \theta_c, \qquad (2.1)$$

where

σ_{os} = interfacial energy between the oil and solid, dynes/cm

σ_{ws} = interfacial energy between the water and solid, dynes/cm

σ_{ow} = interfacial energy (interfacial tension) between the oil and water, dynes/cm

θ_c = angle at the oil-water-solid interface measured through the water, degrees.

Neither the oil-solid nor the water-solid interfacial energies can be measured directly. However, the equivalent terms — the oil-water interfacial tension and the contact angle — can be determined independently in the laboratory.

The contact angle, θ_c, has achieved significance as a measure of wettability. As shown in Fig. 2.1, the value of the contact angle can range from zero to 180° as

BASIC FLOW PROPERTIES OF ROCK

limits. Contact angles of less than 90°, measured through the water phase, indicate preferentially water-wet conditions, whereas contact angles greater than 90° indicate preferentially oil-wet conditions. A contact angle of exactly 90° would indicate that the rock surface has equal preference for water and oil.

References to wettability in a qualitative sense have also appeared. In the technical literature are the terms "strongly water-wet", "strongly oil-wet", or "intermediate wettability". The quantitative range of these qualitative terms is rarely indicated. However the following approximate ranges are sometimes used:

Contact angles in the vicinity of zero and 180° are considered strongly water-wet and strongly oil-wet, respectively. Contact angles near 90° have moderate wetting preference and cover a range termed "intermediate wettability".

In the early stages of reservoir engineering, it was generally considered that all formations are preferentially wet with water. This seemed only natural, considering that sandstone reservoirs were deposited in an aqueous environment and that only later did oil migrate into these sands. Also, most sedimentary rock minerals are water-wet in their natural state. In carbonate formations, water played a large part in the development of porosity, and here, too, oil was a late-comer. However it was recognized by Nutting[10] in 1934 that some producing formations are oil-wet.

Extensive laboratory experience has led to the conclusion that the nature of reservoir wettability is due to the absence or presence of polar compounds existing in minute quantity in crude oil.[11,12] These polar compounds, apparently asphaltic in nature, adsorb on the rock surfaces and tend to make these surfaces oil-wet. The effect of these polar compounds depends to some degree upon the nature of the rock surface—that is, upon whether the rock surfaces are predominantly silica, carbonate, or clay. Denekas et al.[13] made a detailed study of the effect of crude oil components on rock wettability.

In some formations asphaltic materials are so strongly adsorbed that they resist removal by normal core cleaning procedures. Examples of these are the Tensleep sandstone in Wyoming and the Bradford sandstone in Pennsylvania.

Various techniques have been described for determining the wettability of reservoir rock. The principal ones will be discussed.

Contact Angle Measurements

To obtain a representative and true measure of rock wettability by contact angle measurements, an uncontaminated sample of the reservoir crude oil is necessary. The composition of the rock surfaces must be known. In addition a sample of the reservoir water is desirable, though its mineral content can be simulated if no formation water is available. A contact angle measurement basically determines whether the oil contains surfactants that can render reservoir rock oil-wet.

The contact angle cell and the basic test procedure have been described[14] in the literature. A flat polished crystal of the mineral that is predominant in the rock surfaces is immersed in a sample of formation water. A drop of the reservoir oil is then placed on the solid surface. Fig. 2.2 shows how an oil droplet is held between two crystal surfaces. The two plates are moved so that water advances over a portion of the crystal previously covered with oil. The contact angle at the surface newly exposed to water is termed the water-advancing contact angle and is measured as a function of the time the oil has been in contact with the surface.

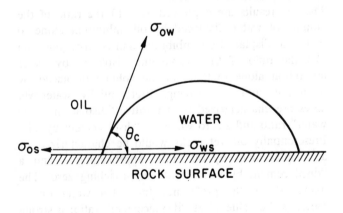

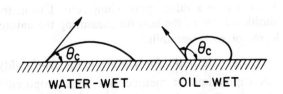

Fig. 2.1 Wettability of oil-water-solid system (after Ref. 8).

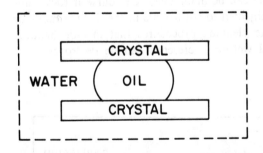

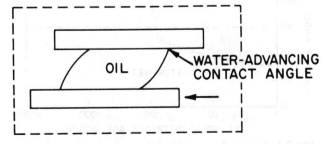

Fig. 2.2 Contact angle measurement.

Fig. 2.3, a typical plot of these measurements, shows that the contact angle increases with the age of the oil-solid interface until an equilibrium value is obtained. Frequently hundreds or even thousands of hours of aging time are required before equilibrium is attained. As shown by Fig. 2.3, early measurements can indicate a water-wetting preference even though at equilibrium the surface is oil-wet.

The main advantages of contact angle measurements are the reliability of the results and the relative ease of obtaining uncontaminated reservoir fluid samples as compared with obtaining uncontaminated rock samples. Disadvantages include the long testing time and the necessary cleanliness and inertness of the test system.

Imbibition-Displacement Tests on Rock Samples

Several types of laboratory tests involving imbibition and displacement procedures with reservoir rock samples have been proposed. In general, these tests use core samples that are handled, shipped, and stored in such a way that their native wettability is preserved. The requirements for coring and for handling cores so that their wettability is not altered will be discussed in Section 2.5.

The tests for evaluating reservoir wettability from core samples frequently involve the use of refined oil and laboratory-prepared brines. One might wonder, then, whether the tests have any bearing upon reservoir wettability since the fluids that have caused that wettability preference have been removed. Actually, the tests are meaningful. As indicated in Fig. 2.3, the reservoir wettability requires hundreds or even thousands of hours of aging time to achieve adsorption equilibrium. Similarly, desorption of these wettability-influencing materials by flow could require corresponding periods of time. As long, therefore, as laboratory tests using refined oil and laboratory-prepared brines are conducted quickly, the wettability preference of the rock sample can be maintained. Of course if rock samples are subjected to surfactants in the drilling muds or to complete cleaning procedures that include heating, the natural wetting preference can be destroyed.

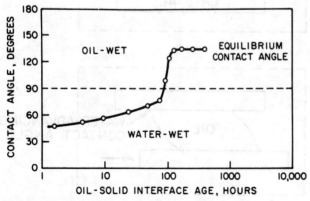

Fig. 2.3 Approach to equilibrium contact angle.

Bobek et al.[15] have proposed a laboratory test to ascertain preferential wettability. It consists of determining which fluid will displace the other from a rock sample by imbibition. The results of this imbibition test are then compared with those of a reference imbibition test on the same core sample after it has been heated to 400°F for 24 hours to remove any organic materials. The assignment of qualitative wettability designations is based on the relative amounts and rates of imbibition in the two tests.

In the same paper there was discussed a method of estimating the wettability of unconsolidated material. This method could be of particular value for well-site testing. A thin layer of the unconsolidated sand is spread on a microscope slide. The oil content of the sand is increased by adding a clear refined oil. Droplets of water are then placed on the surface of the sand grains and the fluid movement is observed. If the sand is water-wet, the added water will displace oil from the surfaces of the sand grains and the oil will form in spherical droplets, indicating that the oil is the non-wetting phase. A similar procedure is used to test for oil wettability.

Amott[16] proposed a combined imbibition-displacement test procedure. After the reservoir core sample has been flushed with water to the residual oil saturation and evacuated to remove gas, it is subjected to the following procedure.

1. It is immersed in oil (e.g., kerosene). The volume of water displaced by the imbition of oil is measured after 20 hours' immersion.
2. It is centrifuged under kerosene and the total volume of water displaced is measured.
3. It is immersed in water and the volume of oil displaced by water imbibition is recorded at the end of 20 hours.
4. It is centrifuged under water and the total oil displaced is recorded.

The test results are expressed as: (1) the ratio of the volume of water displaced by oil imbibition alone to that total displaced by imbibition and centrifuging, and (2) the ratio of the oil volume displaced by water imbibition alone to the total oil volume displaced by imbibition and centrifuging. Preferentially water-wet cores are characterized by a positive "displacement by water" ratio and a zero value for "displacement by oil". Preferentially oil-wet rocks would be characterized by a "displacement by oil" ratio approaching 1 and a "displacement by water" ratio approaching zero. The magnitude of the preference for either water or oil parallels the value of its "displacement" ratio; a strong preference indicated by a value approaching 1, a weak preference by a value approaching zero. This method is considered one of the best for measuring the nature and degree of rock wettability.

Other Methods for Determining Wettability

A capillarimetric method has been proposed[17] for measuring the wetting tendencies of crude oil and water systems. This method consists essentially of determining which fluid, crude oil or water, will displace the other by imbibition from a glass capillary. A major limitation of this method is the assumption that the glass surface of the capillary is representative of the reservoir rock.

Slobod and Blum[18] proposed the use of two semiquantitative terms, "wettability number" and "apparent contact angle", to reflect the wettability of a rock sample. The values of these terms are determined by conducting two displacement tests, the first being the displacement of water by oil and the second being the displacement of oil by air. In each of these experiments the pressure necessary for initial displacement to occur, the water-oil interfacial tension, and the air-oil interfacial tension (i.e., oil surface tension), are all used to calculate the values of "wettability number" and "apparent contact angle". The technical literature reflects little use of this method.

A nuclear magnetic relaxation technique was suggested[19] for determining the portions of the rock surface area that are preferentially water-wet or oil-wet. A rock sample is first exposed to a strong magnetic field, then to a much weaker magnetic field. The magnetic relaxation rate—that is, the rate at which the initially imposed magnetism is lost—is then measured. In specifically prepared sandpacks containing known mixtures of oil-wet and water-wet sand grains, a linear relation was observed between the relaxation rate and the fractional oil-wet surface area. Though the authors reported no studies using natural cores, they proposed a testing procedure. Their technique requires specialized equipment not normally found in petroleum laboratories, and there are no indications in the literature that it has found routine use.

Holbrook and Bernard[20] proposed a method for determining the relative water-wettability of a core sample. This method is based upon observation that a rock surface covered by water and thus considered to be water-wet will adsorb methylene blue (a water-soluble dye) from solution, but an oil-covered surface will not. Comparison of the adsorption capacity of a test specimen with that of an adjacent rock sample extracted using chloroform and methanol yields a measure of the wettability. All reported tests were on artificial systems, and the applicability of this technique to natural rocks has not yet been demonstrated.

As early as 1939 it was demonstrated[21] that rock wettability affects oil displacement. Subsequently, laboratory-measured relative permeabilities[22] were found to reflect differences in wettability. (The effect of wettability upon relative permeability will be discussed in Section 2.4.) It is important at this time to gain an understanding of how wettability controls fluid distributions within rock pore spaces.

2.2 Fluid Distribution

An understanding of the distribution of oil, gas, and water within rock pore spaces was limited at first to inferences drawn from the results of laboratory flow tests.

In 1949 and 1950 two efforts were started that were to result in a definitive study of fluid distribution in pore space and its change with flooding history.

The first of these was the American Petroleum Institute Research Project 47B, set up at The U. of Oklahoma.[23] Microscopic studies of dynamic fluid behavior were made in synthetic porous matrices. Basically these porous matrices consisted of a single layer of spheres sandwiched between two transparent flat plates. The fluids used were water and a filtered crude oil. Simultaneous flow of oil and water through these cells was observed and photographed, and the results appeared as a motion picture that received wide distribution and acclaim.

The photomicrographs showed that the water and oil moved in what is termed "channel flow conditions". That is, each fluid moved through its own network of interconnecting channels. The channels varied in diameter from about one grain diameter to many. They were bounded by liquid-liquid interfaces as well as by liquid-solid interfaces and meandered tortuously through the flow cell.

With a change in saturation the geometries of the flow channels were altered. As the oil saturation increased there was a general increase in the number of oil-carrying channels and a corresponding reduction in the number of water channels. A tendency was noted for the channels to hold their positions in the flow beds. It was also observed that the flow within any channel was streamline and devoid of eddy currents, despite the tortuosities in the flow paths.

The residual oil saturations following waterflooding were also observed. The most apparent residual oil saturations existed in large volumes continuous over many grain diameters. Almost always, smaller patches of residual oil were found in the bed.

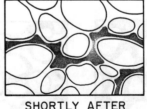
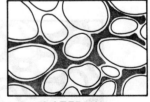

SHORTLY AFTER BREAKTHROUGH   LATER IN FLOOD LIFE

LEGEND
○ SAND GRAIN
□ WETTING FLUID
■ NON-WETTING FLUID

Fig. 2.4 Channel flow concept of fluid flow — nonwetting fluid displacing wetting fluid (after Ref. 24).

The other study of fluid distributions in porous materials was made by Amoco Production Co.[24] (formerly Stanolind Oil and Gas Co.). Sand grains were carefully packed in a cylindrical tube. The wetting phase was simulated by Wood's metal. The nonwetting phase was represented by colored plastic. At any saturation condition the Wood's metal and colored plastic were solidified in place in the core. The core face was magnified and photographed as the face was cut away, giving a graphic, three-dimensional effect when the photographs were projected at motion picture speed.

Fig. 2.4 shows two drawings depicting channel flow at different stages of flooding. Each fluid, wetting and nonwetting, moves in its own network of pores, but with some wetting fluid in each pore. As the nonwetting phase saturation increases, more of the pores are nearly filled with the nonwetting fluid.

Fig. 2.5 illustrates the fluid distribution during a waterflood of a preferentially water-wet formation. In the unaffected portion of the reservoir, the water saturation (connate water) is low and exists as a film around the sand grains and in the re-entrant angles. The remaining pore space is full of oil. In the zone in which water and oil both are flowing, part of the oil exists in continuous channels, some of which have dead-end branches. Other oil has been isolated and trapped as globules by the invasion of water. At floodout, only trapped, isolated oil exists in the rock.

Fig. 2.6 shows a similar history during a waterflood of an oil-wet, initially oil-saturated rock. As the nonwetting phase (water in this case) enters the rock it first forms tortuous but continuous flow channels through the largest pores. As water injection continues, successively smaller pores are invaded and join to form other continuous channels. When sufficient flow channels form to permit almost unrestricted water flow, oil flow practically ceases. The residual oil saturation exists in the smaller flow channels and as a film in the larger, water-filled flow channels.

Figs. 2.5 and 2.6 show that the distribution of either the wetting or nonwetting phase within the pore spaces does not depend solely upon the saturation of that phase, but depends also upon the direction of the saturation change. We should define, now, two terms that

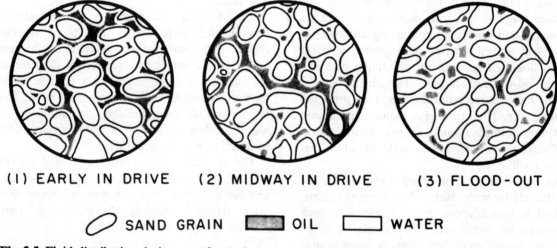

Fig. 2.5 Fluid distribution during waterflood of water-wet rock (after Ref. 24).

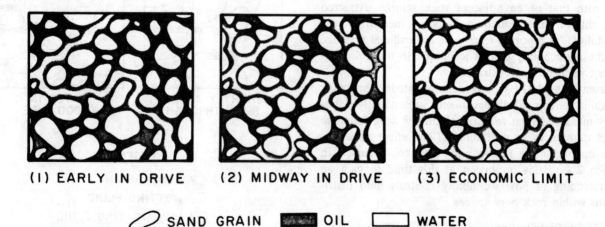

Fig. 2.6 Fluid distribution during waterflood of oil-wet rock (after Ref. 24).

BASIC FLOW PROPERTIES OF ROCK

indicate the direction of the saturation change. "Drainage" refers to flow resulting in a decrease in the wetting phase saturation; the term "imbibition" refers to flow resulting in an increase in the wetting phase saturation. For example, waterflooding a preferentially water-wet rock is an imbibition process, but waterflooding an oil-wet rock is a drainage process.

Since the wettability and direction of saturation change influence the fluid distribution, we would expect these factors to affect similarly both capillary pressure and relative permeability characteristics.

2.3 Capillary Pressure

The first exposure of most engineers to the concept of capillary pressure comes upon witnessing the standard physics experiment in which a capillary tube is inserted into a container of water, and the water rises in the tube. Calhoun gives a good discussion of these experiments and from them draws the concept of capillary pressure.[3] At this point let us define capillary pressure in porous media simply as the pressure difference existing across the interface separating two immiscible fluids, one of which wets the surfaces of the rock in preference to the other.

To a scientist the capillary pressure generally is expressed as the pressure in the nonwetting phase minus the pressure in the wetting phase, and so is commonly a positive value. We have seen in Section 2.1 that rocks can be either preferentially oil-wet or preferentially water-wet. We will define the water-oil capillary pressure as the pressure in the oil phase minus the pressure in the water phase, or

$$P_c = p_o - p_w \quad \ldots \ldots \ldots \ldots (2.2)$$

Thus the capillary pressure can have either a positive or a negative value, depending upon the wettability preference. In discussing gas-water capillary pressure, we will define it as the pressure in the gas phase minus the pressure in the water phase, or

$$P_c = p_g - p_w \quad \ldots \ldots \ldots \ldots (2.3)$$

In Section 2.2, we found that the fluid distribution within the rock pore spaces of given wettability depends upon the direction of the saturation change. The term "hysteresis" is applied to the difference in multiphase rock properties that depends upon the direction of saturation change.

In his pioneering work in 1941, Leverett[25] presented data on the capillary pressure characteristics of unconsolidated sand during the drainage of water as well as during imbibition. Other early investigators[4,26-28] presented data on both drainage and imbibition capillary pressure characteristics, as well as discussions on their significance. An idealized schematic of the change in fluid distribution during the drainage and imbibition processes is shown in Figs. 2.7 and 2.8, respectively.

Perhaps the most thorough study of oil-water capillary pressure characteristics was presented by Killins et al.[29] Oil-water capillary pressures were determined on consolidated sandstones, under both water-wet and oil-wet conditions. Figs. 2.9 through 2.12 show some of the results presented in their study.

Fig. 2.9 shows the drainage and imbibition capillary pressure characteristics of a strongly water-wet sample of Venango sandstone. Note that the pressure in the oil phase must exceed that in the water phase before oil will enter the initially water-saturated rock. This entrance pressure is commonly referred to as the "threshold pressure" or "displacement pressure". Its value is a measure of the degree of the rock wettability, the oil-water interfacial tension, and the diameter of the largest pore on the exterior of the rock sample. A high displacement pressure indicates either a strong degree of wetting or small pores, or both. The slope of the capillary pressure curve during drainage is a good qualitative measure of the range of the pore size distribution. The more nearly horizontal or the flatter the capillary pressure curve, the more uniform the pore sizes within the rock. For comparison, refer to Fig. 2.10, which shows the drainage capillary pressure characteristics for a closely sized glass-bead pack[30]; the imbibition capillary

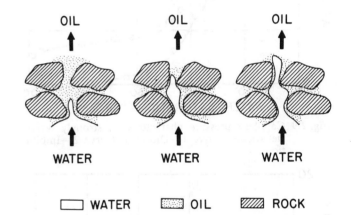

Fig. 2.7 Schematic of drainage process — oil displacement by water, oil-wet sand, $\theta_c = 180°$ (after Ref. 8).

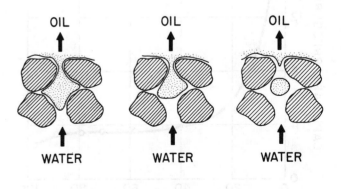

Fig. 2.8 Schematic of imbibition process — oil displacement by water, water-wet sand, $\theta_c = 0°$ (after Ref. 8).

pressure curve is markedly flatter than that for Venango consolidated sandstone (Fig. 2.9).

At the end of the drainage cycle, the core samples were allowed to imbibe the wetting phase. Note the hysteresis in the capillary pressure curves. The imbibition curve shown on Fig. 2.9 reflects a water saturation at the end of imbibition of about 78 percent pore space.

Fig. 2.11 shows the drainage and imbibition capillary

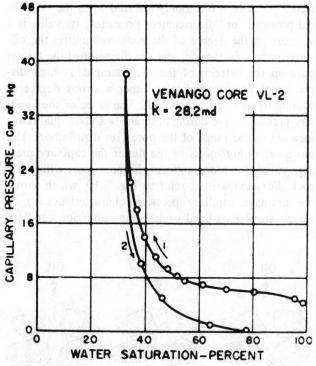

Fig. 2.9 Capillary pressure characteristics, strongly water-wet rock.[29] Curve 1—Drainage. Curve 2—Imbibition.

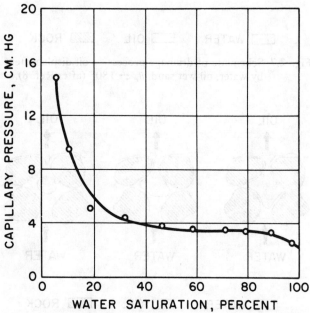

Fig. 2.10 Drainage capillary pressure characteristics (after Ref. 30).

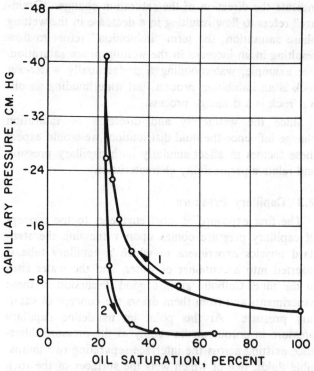

Fig. 2.11 Oil-water capillary pressure characteristics, Tensleep sandstone, oil-wet rock (after Ref. 29). Curve 1 — Drainage. Curve 2 — Imbibition.

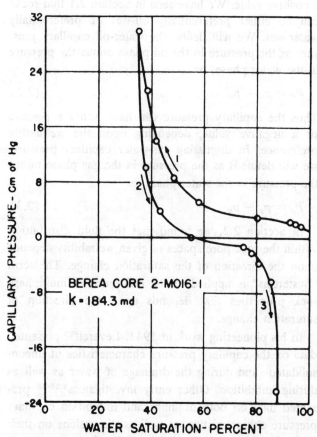

Fig. 2.12 Oil-water capillary pressure characteristics, intermediate wettability.[29] Curve 1—Drainage. Curve 2 — Spontaneous Imbibition. Curve 3 — Forced Imbibition.

BASIC FLOW PROPERTIES OF ROCK

pressure curves for a strongly oil-wet sample of Tensleep sandstone. Though the vertical coordinates have a reversed sign to account for the difference in wettability, note the similarities in the drainage and imbibition characteristics shown in Figs. 2.9 and 2.11.

To compare with the characteristics of rock samples having a strong wettability preference, note Fig. 2.12, which shows the capillary pressure characteristics of a sample of Berea sandstone having intermediate wettability. We can deduce from the drainage capillary pressure data that the sample is moderately water-wet. This is indicated by the small positive value of the threshold capillary pressure, or displacement pressure. At the end of the drainage cycle (Curve 1), the rock sample spontaneously imbibed (to zero capillary pressure), reaching a 55 percent water saturation (Curve 2). As positive water pressures were applied, the water saturation increased (at positive capillary pressures) until a maximum water saturation of about 88 percent was attained (Curve 3). At this point further increases in water pressure resulted in no further saturation change.

Unfortunately, water-oil capillary pressure data are difficult to measure and thus too rarely are obtained. The more commonly measured capillary pressure characteristics are those for the air-brine system. In these measurements, the core sample cleaned of reservoir fluids is completely saturated with brine, and air is used to desaturate the rock of water. A typical air-brine capillary pressure curve is shown in Fig. 2.13.

Many engineers use the air-brine capillary pressure curve to provide an estimate of the reservoir connate water saturation. They reason that the oil-bearing formations were initially saturated with water and then oil migrated in to displace the water. This process then, they continue to reason, can be simulated by an air-brine capillary pressure curve. The minimum water saturation obtained from the capillary pressure curve, or the saturation at the capillary pressure corresponding to the height above the water-oil contact, is taken to be the reservoir connate water saturation.

More frequently than not, the use of air-brine capillary pressure curves to estimate reservoir connate water saturations is misleading. It implicitly assumes that the reservoir rock is wet by water in the presence of oil to the same degree as the cleaned sample of reservoir rock is wet by water in the presence of air. Obviously this assumption is erroneous in the case of preferentially oil-wet and intermediate-wettability rocks, since brine strongly wets most cleaned rock in the presence of air. Therefore air-brine capillary pressure curves are useful only in formations known to be strongly water-wet.

Another form of capillary pressure curve frequently measured is that resulting from the injection of mercury into a clean, dry rock sample.[4] Since mercury does not wet the rock surfaces, a drainage-type curve is obtained. The advantage of using the mercury injection technique is that irregularly shaped rock samples — perhaps bit cuttings or sidewall cores — can be tested. Fig. 2.14 shows an air-mercury capillary pressure curve for a well sorted sand. As a recent paper[31] pointed out, air-mercury capillary pressure data reflect the fluid distribution in water-oil systems only under strong wetting conditions. Brown[32] found that gas-oil capillary pressure data would agree with mercury injection capillary pressure data if an appropriate scaling factor is used.

Leverett's capillary pressure function[25] was proposed by Rose and Bruce[33] for correlating capillary pressure data. Since these data reflected the pore size distribution, the radius of the largest pore, the wettability, and the interfacial tension of the fluid pair involved, it should be possible to normalize it by a so-called J-function, where

$$J(S_w) = \frac{P_c}{\sigma_{ow} \cos \theta_c} \sqrt{\frac{k}{\phi}} \quad \ldots \ldots \quad (2.4)$$

Brown[32] found that the use of this J-function was successful in correlating capillary pressure data on a number of core samples from the Edwards limestone in the Jourdanton field, southwest Texas. However, Brown implied that its usefulness was limited to specific lithologic types within the same formation.

2.4 Relative Permeability

The relative permeability characteristics are a direct measure of the ability of the porous system to conduct

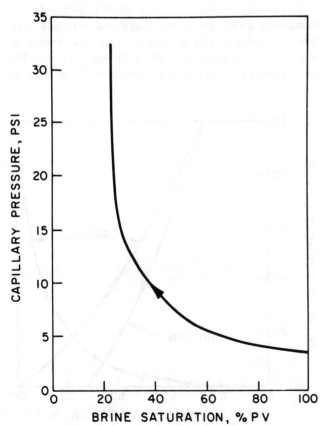

Fig. 2.13 Air-brine drainage capillary pressure characteristics.

one fluid when one or more fluids are present. These flow properties are the composite effect of pore geometry, wettability, fluid distribution, and saturation history.

Relative permeability was earlier defined as the effective permeability to a specific fluid divided by some base permeability. Three different base permeabilities are generally used: (1) the absolute air permeability, (2) the absolute water permeability, and (3) the permeability to oil at reservoir connate water saturation. The third base will be used throughout this Monograph, and as a consequence, the relative permeability to oil at connate water saturation is 1.0, or 100 percent.

Because the saturation history affects the fluid distribution and causes a hysteresis in the capillary pressure characteristics, we should expect to see a similar hysteresis effect in the drainage and imbibition relative permeability characteristics. Fig. 2.15 shows a typical set of drainage and imbibition relative permeability characteristics in terms of the wetting phase saturation. Note that the wetting phase relative permeability is a function only of its own saturation. That is, during imbibition the wetting phase permeabilities retrace those obtained during drainage to the maximum wetting phase saturation (i.e., the saturation corresponding to a zero oil permeability). This is observed in systems with strong wettability preference. The nonwetting phase, however, has a lower relative permeability at any saturation during imbibition than it does during drainage.

Typical water-oil relative permeability characteristics are presented for water-wet and for oil-wet formations in Figs. 2.16A and 2.17A, respectively; they appear on arithmetic coordinates. Figs. 2.16B and 2.17B show the same flow properties on semilog coordinates. The differences in the flow properties that indicate the different wettability preferences can be illustrated by the following rules of thumb:

	Water-Wet	Oil-Wet
Connate water saturation	Usually greater than 20 to 25 percent PV	Generally less than 15 percent PV, frequently less than 10 percent
Saturation at which oil and water relative permeabilities are equal	Greater than 50 percent water saturation	Less than 50 percent water saturation
Relative permeability to water at maximum water saturation; i.e., floodout	Generally less than 30 percent	Greater than 50 percent and approaching 100 percent

It has generally been found that rocks with intermediate wettability have some of the above characteristics of both water-wet and oil-wet formations. It is the experience of those who have measured a number of water-oil flow properties that most formations are of intermediate wettability—that is, with no strong preference for either oil or water.

One highly important factor should always be remembered about laboratory-determined water-oil relative permeability characteristics. The flow properties are those of the formation *only* if the sample's wettability preference is that of the formation. Thus, vital in

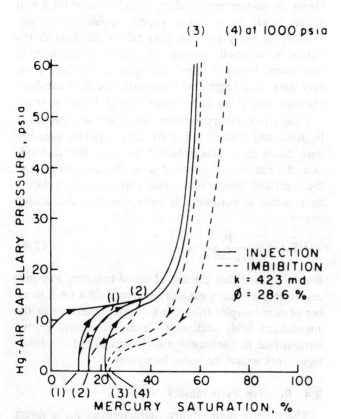

Fig. 2.14 Air-mercury capillary pressure characteristics, well sorted sand.[31]

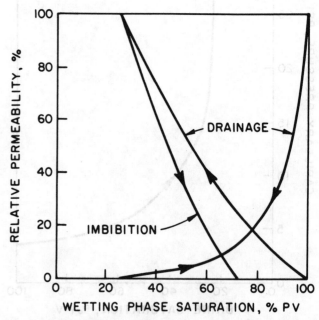

Fig. 2.15 Drainage and imbibition relative permeability characteristics.

BASIC FLOW PROPERTIES OF ROCK

obtaining representative relative permeabilities is the proper handling of core samples to insure that wettability is maintained from the formation to the laboratory test apparatus.

2.5 Native-State and Restored-State Cores

The term "native-state" as used in this section implies that the rock sample retains the native wettability preference and frequently the reservoir fluid saturations from the formation to the laboratory. "Restored-state" implies that the core sample is cleaned and dried and, following this, its wettability and fluid saturation are restored to those conditions thought to exist in the reservoir. Obviously to restore a core sample to reservoir wettability and saturation requires an independent measurement of these two properties. On the other

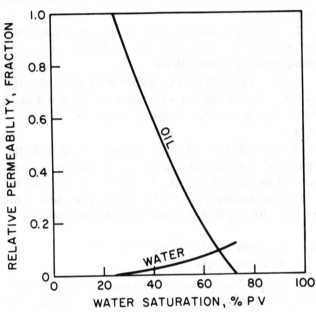

Fig. 2.16A Typical water-oil relative permeability characteristics, strongly water-wet rock.

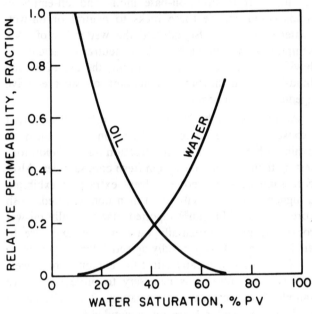

Fig. 2.17A Typical water-oil relative permeability characteristics, strongly oil-wet rock.

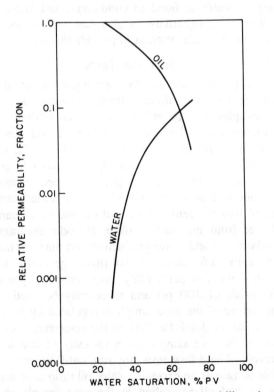

Fig. 2.16B Typical water-oil relative permeability characteristics, strongly water-wet rock.

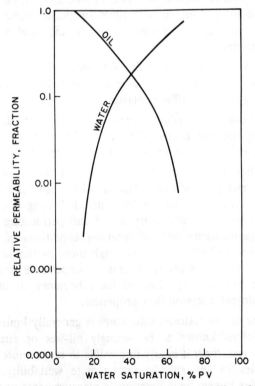

Fig. 2.17B Typical water-oil relative permeability characteristics, strongly oil-wet rock.

hand, if it can be assured that the coring, handling, transporting, and storage procedures used in getting core samples to the laboratory do not alter their wettability, the independent information on formation wettabiltiy is not necessary. Thus native-state cores, properly handled, are the more desirable.

In 1958, Bobek et al.[15] presented a detailed study of the effect that coring fluids and core-handling procedures have on alteration of rock wettability. They found that prolonged exposure of naturally water-wet rocks to filtrate from both oil-base muds and oil-emulsion muds could change these rocks to neutral or oil-wet. Water-base muds also affected the wettability of rock samples, with bentonitic muds of neutral or acidic pH having the least effect. In any event, the use of a low-fluid-loss mud having a minimum of surface-active agents is recommended.

An early study by Richardson et al.[34] indicated that exposure of core samples to air can cause them to exhibit a higher residual oil saturation during laboratory testing than that obtained from fresh core samples. Also, core samples subjected to solvent extraction exhibited a higher residual oil saturation than nonextracted, fresh core samples. Mungan[35] reported that naturally oil-wet rocks become preferentially water-wet by exposure to air for a week. It is generally agreed[15] that exposure to air (oxygen) can greatly affect the exhibited rock wettability. Therefore, it is necessary to protect the core sample from oxygen contamination. Two alternative packaging procedures are recommended.[15]

1. Immerse the core material at the well site in either deoxygenated formation brine or synthetic brine contained in glass-lined steel tubes or plastic tubes that can be sealed against both content leakage and oxygen penetration.

2. Wrap the cores at the well site in Saran or polyethylene film, covered by aluminum foil. Then coat this package with paraffin or plastic.

By either procedure, the cores are prevented from drying out and protected from oxygen. Because of its relative ease, the latter procedure is preferable.

Other investigators have commented on the laboratory testing conditions needed with these preserved cores. Recommendations have included using reservoir crude oil in the laboratory tests[36] and performing relative permeability tests at reservoir conditions of temperature and pressure.[37] Although these particular precautions are not always necessary, experience and extreme care are required in the laboratory to obtain meaningful water-oil flow properties.

The use of restored-state cores is generally limited to formations known to be strongly oil-wet or strongly water-wet, for it is nearly impossible to reconstitute core samples to any specific intermediate wettability. For samples known by contact angle measurements or well-site imbibition tests to be strongly water-wet, the recommended procedure is to (1) saturate the core sample with water, (2) displace a portion of this water with refined oil until a simulated reservoir connate water saturation is obtained, and (3) perform the relative permeability test.

Extensive laboratory experience has shown that the reservoir connate water in oil-wet rocks exists in tiny droplets in the smaller pores of the rock. These water globules play no part in the displacement of oil by injected water. These conclusions were drawn from the following observations from core tests on native-state oil-wet rock:

1. The oil permeability with connate water saturation present closely approximates that with the core sample completely saturated with oil.

2. The water-oil relative permeability characteristics at low values of injected water saturation do not depend upon whether connate water saturation is present or absent.

It is impossible at this time to reconstitute both the magnitude and distribution of the connate water saturation in an oil-wet rock. Therefore, for these rocks, the recommended procedure is to saturate them completely with oil, then perform the water-oil flow test.

2.6 Methods of Measuring Water-Oil Flow Properties

Water-oil flow properties of reservoir rock are generally used to provide an estimate of the oil recovery that might be obtained by flushing a unit volume of the reservoir with water. Information on the relative permeability to water at floodout conditions, and hence the possible water injectivity, is also desirable. Flood pot tests were the initial attempt to provide this information.

Flood Pot Tests

Flood pot tests are usually carried out on cylindrical whole core samples flooded from the inside out. The core samples may be either cut with an oil-base mud or resaturated to the reservoir water and oil content. The normal procedure[38] is to drill a hole, ¼ in. or so in diameter, at the central axis of the cylinder to about 90 percent of the core's length. The core sample is then mounted in a "flood pot" so that water can be injected into the central hole and oil and water can be collected from the outer surface. Periodic readings of cumulative oil and water production and time are made. A pressure differential of 10 psi is generally used, though with low-permeability core samples, pressure differentials of 100 psi and more may be used. The temperature of the core sample is regulated by an electric heater to duplicate that in the reservoir; thus the crude in the core sample has a viscosity of that in the reservoir. These flow tests are run until either a negligible oil rate or an excessive water-oil ratio is obtained. The residual oil saturation in the core sample at the end of the test is determined from extraction of the whole sample or a small core plug taken from it.

From the test results the residual oil saturation at floodout can be determined, as well as the volume of injected water necessary to obtain the corresponding oil recovery. Additional information frequently obtained includes the pressure differential required to effect displacement, the existence of an oil bank, the relative permeability to oil in the oil bank, and the relative permeability to water in the flooded-out region of the reservoir.[38]

As the technology has grown it has become apparent that information limited to the residual oil saturation will not suffice for good engineering predictions of waterflood performance. The increased use of the Buckley-Leverett[39] frontal advance theory has increased the need for a knowledge of the entire relative permeability-saturation relationships.

Steady-State Tests

One of the most important properties needed to engineer a waterflood is the water-oil relative permeability characteristics of the reservoir rock. Relative permeability characteristics, when properly determined, measure the composite effect of pore geometry, wettability, fluid distribution, and saturation, as well as saturation history.

In the late 1940's and early 1950's, a number of research organizations made a determined effort to ascertain the influence of many factors upon relative permeability measurements. The objective was to develop a simple and reliable experimental technique for obtaining these flow properties. Of interest were the gas-oil relative permeabilities as well as the relative permeabilities to water and oil. For completeness, therefore, we shall discuss briefly the general subject of relative permeability testing techniques.

The measurement of relative permeability basically involves determining the flow rate of oil, water, or gas at a known fluid saturation and a specified pressure differential. From these measurements, the fluid relative permeabilities can be calculated.

Early experimenters suggested techniques that would fix the fluid saturation and measure the corresponding relative permeabilities. These techniques included the Hassler method,[40] the Penn State method,[41] the stationary liquid procedure,[42] the single core dynamic,[43] the intermittent gas drive,[44] the Hafford method,[45] and the dispersed feed procedure.[45] Many of these methods are closely related and indicate attempts to eliminate both inlet and outlet (boundary) end effects, which manifest themselves in saturation gradients. Richardson et al.[45] reported that boundary effects could be eliminated and that the Penn State, Hafford, Hassler, and dispersed feed techniques gave valid results. Outlet end effects occurring during their experiments were measured[45,46] and it was found that they agreed in magnitude with those predicted by Elkins.[47] Geffen et al.[22] found that if the pressure gradients across the test sample were increased, boundary effects could be eliminated. It was further observed[45] that drainage relative permeability characteristics were independent of flow rate as long as the rate was not so high as to cause inertial effects. Other experiments[48] confirmed that laboratory-measured relative permeabilities are independent of flow rate, provided there is no saturation gradient induced in the core sample by boundary effects. Furthermore, relative permeability characteristics were found to be independent of fluid viscosities.[22] However, since the relative permeabilities depend upon the direction of saturation change, the direction of change in the laboratory testing must correspond to that in the reservoir.

The common descendant from many of these techniques is the so-called steady-state method. In this method, water and oil (in the case of water-oil flow property determinations) are injected simultaneously at a fixed ratio. The pressure differential during flow is measured and from this the water and oil relative permeabilities are determined. After the fluid saturations within the test core have reached an equilibrium, they are determined either by weighing or by electrical resistivity measurements.

External-Drive Techniques

In 1952, Welge[49] presented a new technique for determining the relative permeability characteristics of a rock sample from its performance during an external gas or water drive. Using the Buckley-Leverett frontal advance equation,[39] Welge developed the following relationship:

$$\overline{S}_d - S_{d2} = Q_i f_{o2} , \qquad \ldots \ldots \ldots (2.5)$$

where

$\overline{S}_d =$ average displacing phase saturation, fraction PV

$S_{d2} =$ displacing phase saturation at the exit end of the test sample, fraction PV

$Q_i =$ pore volumes of cumulative injected fluid, dimensionless

$f_{o2} =$ fraction of oil in the produced stream, fraction

In this equation, the difference between the average and producing end saturation is related to the cumulative injected fluid and the producing oil cut. In practice, the values of the terms $\overline{S}_d$, Q_i, and f_{o2} can be determined at any time from the production history of a flow experiment. Then the saturation at the outflow face, S_{d2}, can be calculated from Eq. 2.5. Also, the producing water-oil ratio (WOR) or the gas-oil ratio (GOR) and the corresponding relative permeability ratio can be determined from the experimental data. This relative permeability ratio can then be related to the corresponding outflow face saturation. A limitation of this method at the time it was introduced was that it could yield only the relative permeability ratio, not the individual relative permeabilities.

Stewart et al.[50] showed that Welge's procedure applied to an external gas drive gave the same relative permeability characteristics as obtained during a solution gas drive in which there was no saturation gradient. Owens et al.[51] showed that relative permeability ratios determined by the Welge technique agreed with those obtained from steady-state tests.

In 1959, Johnson et al.[52] presented a mathematical technique for obtaining individual relative permeabilities from external-drive laboratory tests. As verification of their method, the authors showed excellent agreement between relative permeabilities calculated in this manner and those determined on the same rock samples using the steady-state technique.

The advantage of the external-drive technique is its simplicity and its speed. However, it has a severe limitation for determining water-oil flow properties. This limitation arises from two conditions normally imposed on the testing: (1) pressure differentials across the test core during testing are frequently 50 psi and above to eliminate outlet end effects;[53] and (2) viscous oils are generally used so that relative permeabilities over the maximum saturation range can be obtained. As a consequence of these two conditions, preferentially water-wet porous media frequently behave as if they were oil-wet during an external water drive.[53] The displacement occurs at such a speed that with viscous oil filling the pores initially the natural water wettability of the rock existing at equilibrium conditions could not be manifest. There is no universal remedy. The external-drive technique for measuring water-oil flow properties is not recommended for rocks known to be preferentially water-wet. If no information on the natural wettability is available, the relative permeabilities thus obtained should be used only with caution.

Throughout the years there have been a number of discussions on the effect of the water-oil interfacial tension on the laboratory measured flow properties of rock.[54,55] The general consensus is that interfacial tensions must be reduced to 1.1 dynes/cm[56] or lower[57] before significant reductions in residual oil saturation are obtained. For reference, the normal water-oil interfacial tension ranges from about 23 to 30 dynes/cm, far above the level where residual oil saturations can be affected.

Fatt[58] has also determined that overburden pressure has no effect on laboratory-measured relative permeability characteristics. The only effect is on the specific permeability. McLatchie et al.[59] have presented data on the effect of overburden pressure on the permeability of sandstones.

Determination of Relative Permeability by Other Means

Burdine[5] first proposed the use of pore size distribution to calculate the relative permeability characteristics of reservoir rock. Others[60-64] have worked in this same area, studying both the capillary pressure and relative permeability characteristics of interconnected capillaries and finding them similar to those of reservoir rock. Little success has been found in applying pore size distribution data to calculation of relative permeability characteristics of reservoir rock because this technique requires porous systems having strong wetting properties that are known. Likewise, calculation of water-oil flow properties from mercury injection capillary pressure data has been proposed but appears to be limited to rock with strong wettability preference.

It has been proposed[65] that relative permeabilities be calculated from electrical resistivity measurements, but this also has had limited utility.

Averaging Relative Permeability Characteristics

Every practicing reservoir engineer has been faced with a number of laboratory-determined relative permeability characteristics, all different, for the one formation of interest. Sometimes he finds that the flow properties appear to be related to the sample permeability. Figs. 2.18 and 2.19 show what might be encountered for preferentially oil-wet and water-wet formations, respectively. In this event, the engineer's problem is simplified. By interpolating between these curves, he may come up with the relative permeability characteristics for the average formation permeability. On the other hand, if a study of the reservoir heterogeneity indicates that the reservoir is composed of discrete layers or strata, each of which has a characteristic permeability, water-oil flow properties for each layer can be obtained by interpolation.

More likely than not the orderly arrangement of water-oil flow properties shown in Figs. 2.18 and 2.19 is not obtained. When this is the case and an average of a number of relative permeability curves is required, the following procedure[46] is recommended:

> The relative permeability curves are presented on a semilogarithmic plot rather than on the conventional Cartesian (arithmetic) coordinates. Average saturation values are determined arithmetically at equal values of relative permeabilities and permeability ratios. The average curves are replotted on the basis of total porosity, using average values for connate water and residual oil saturations.

2.7 Three-Phase Relative Permeabilities

Several technical papers have been published in which three-phase relative permeability measurements are discussed[61,66-69] and data are presented.[70] Such data are of value for calculating the performance of processes in which three phases — water, oil, and gas — flow simultaneously through the rock pore spaces. However, in conventional waterflooding operations the portion of the reservoir in which all three phases flow is insignificant, even when the gas saturation before waterflooding is large. This will be discussed in greater detail in Section 3.9.

2.8 Connate Water Saturation

The connate water saturation is by definition the water saturation existing in the reservoir at discovery. It is generally but not always true that the connate water saturation is so low that it has no permeability; that is, water does not flow upon production. The value of connate water saturation determines by difference the volume of reservoir oil in place.

When wells are drilled using a water-base mud, the water filtrate serves to increase the water saturation in the formation adjoining the hole as well as in any rock sample being cored. This complicates connate water evaluation by well logging techniques. Cores cut with water-base muds cannot give a reliable estimate of the reservoir connate water saturation. On the other hand, cores cut with an oil-filtrate mud can give an accurate estimate of connate water saturation for those formations containing immobile connate water. Another reliable technique for obtaining reservoir water saturations utilizes gas-cut cores. Any evaporation of water contained in the core sample by the gas is generally insignificant.

Another technique, termed the evaporation method,[71] has been proposed for determining the reservoir connate water saturation. This method is limited to water-wet rocks, however, and even in this case determines not the connate saturation but the maximum saturation at which water flow does not occur.

In performing meaningful laboratory relative permeability tests, the magnitude of the reservoir connate water saturation is important. Laboratory experience has shown that the connate water saturation in preferentially oil-wet cores has no effect on the relative permeabilities as long as the connate saturation is less than about 20 percent PV. In preferentially water-wet rock, the initial water saturation has a definite effect upon the measured water-oil relative permeability characteristics (Fig. 2.20). Therefore, in these rocks the interstitial water saturation at the start of testing should closely approximate the reservoir connate water saturation.

Most engineers have encountered the situation in which the water-oil relative permeability curves for the reservoir of concern have a different irreducible connate water saturation than that chosen as the average reservoir value. Should the measured water-oil flow properties be used directly or should they be adjusted in some way to account for the differences in connate water saturation? The recommended procedure[72] calls for a

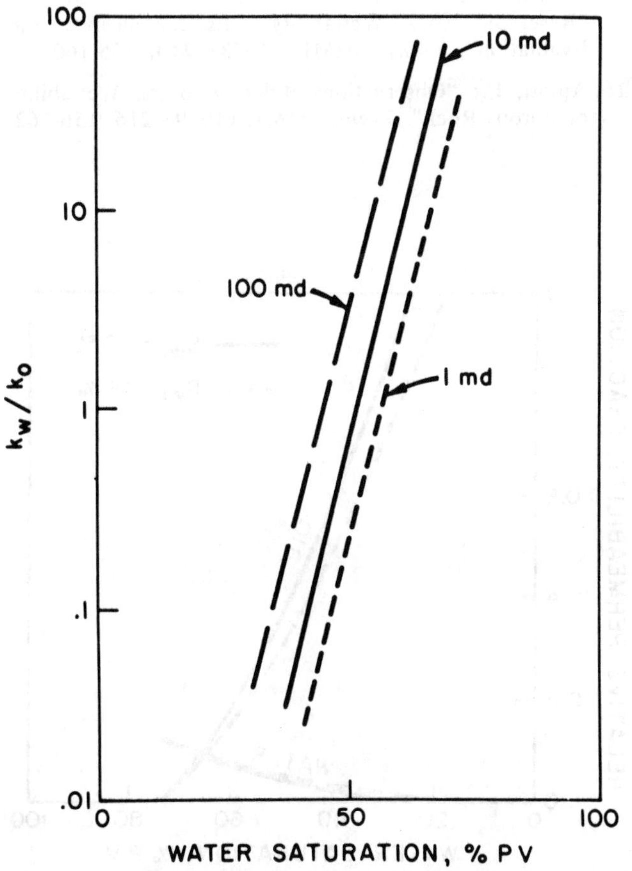

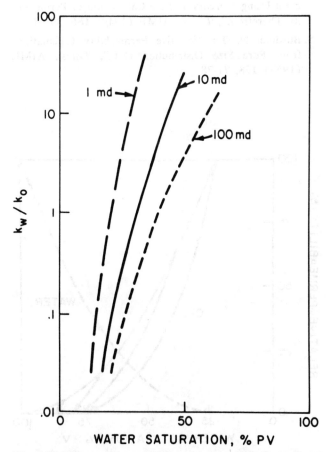

Fig. 2.18 Effect of permeability on water-oil flow properties, preferentially water-wet rock.

Fig. 2.19 Effect of permeability on water-oil flow properties, preferentially oil-wet rock.

new set of water-oil relative permeability curves, constructed so that they adhere to the following criteria.

1. The relative permeability to oil is 1.0 and to water is zero at the *reservoir* connate water saturation.

2. The relative permeability to water at floodout and the water saturation at this point are the same as in the *laboratory* test results (the relative permeability to oil at floodout being zero, of course).

3. The shape of the individual "reservoir" relative permeability curves is similar to that of the laboratory-developed curves between the two end points of each curve.

Fig. 2.21 is an example of the application of this technique. Illustrated are the relative permeability data shown in Fig. 2.16 for a connate water saturation of 25 percent PV and the adjusted data for a connate water saturation of 18 percent.

References

1. Muskat, M.: *Physical Principles of Oil Production*, McGraw-Hill Book Co., Inc., New York (1949).
2. Pirson, S. J.: *Elements of Oil Reservoir Engineering*, McGraw-Hill Book Co., Inc., New York (1950).
3. Calhoun, J. C., Jr.: *Fundamentals of Reservoir Engineering*, The U. of Oklahoma Press, Norman (1960).
4. Purcell, W. R.: "Capillary Pressures — Their Measurement Using Mercury and the Calculation of Permeability Therefrom", *Trans.*, AIME (1949) **186**, 39-48.
5. Burdine, N. T.: "Relative Permeability Calculations from Pore Size Distribution Data", *Trans.*, AIME (1953) **198**, 71-78.
6. Corey, A. T.: "The Interrelation Between Gas and Oil Relative Permeabilities", *Prod. Monthly* (Nov., 1954) **19**, No. 11, 34-41.
7. Hough, E. W., Rzasa, M. J. and Wood, B. B., Jr.: "Interfacial Tensions at Reservoir Pressures and Temperatures, Apparatus and the Water-Methane System", *Trans.*, AIME (1950) **192**, 57-62.
8. Raza, S. H., Treiber, L. E., and Archer, D. L.: "Wettability of Reservoir Rocks and Its Evaluation", *Prod. Monthly* (April, 1968) **33**, No. 4, 2-7.
9. Adams, N. K.: *The Physics and Chemistry of Surfaces*, Oxford U. Press, London (1941).
10. Nutting, P. G.: "Some Physical and Chemical Properties of Reservoir Rocks Bearing on the Accumulation and Discharge of Oil", *Problems in Petroleum Geology*, AAPG (1934).
11. Benner, F. C. and Bartell, F. E.: "The Effect of Polar Impurities Upon Capillary and Surface Phenomena in Petroleum Production", *Drill. and Prod. Prac.*, API (1941).
12. Leach, R. O.: "Surface Equilibrium in Contact Angle Measurements", paper presented at Gordon Research Conference on Chemistry at Interfaces, Meridan, N. H. (July, 1957).
13. Denekas, M. O., Mattax, C. C. and Davis, G. T.: "Effects of Crude Oil Components on Rock Wettability", *Trans.*, AIME (1959) **216**, 330-333.
14. Wagner, O. R. and Leach, R. O.: "Improving Oil Displacement Efficiency by Wettability Adjustment", *Trans.*, AIME (1959) **216**, 65-72.
15. Bobek, J. E., Mattax, C. C. and Denekas, M. O.: "Reservoir Rock Wettability — Its Significance and Evaluation", *Trans.*, AIME (1958) **213**, 155-160.
16. Amott, E.: "Observations Relating to the Wettability of Porous Rock", *Trans.*, AIME (1959) **216**, 156-162.

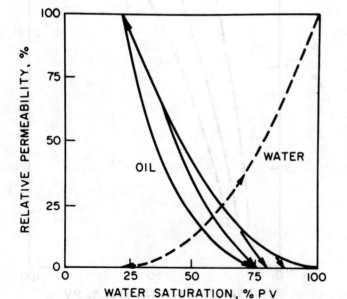

Fig. 2.20 Effect of initial water saturation on water-oil flow properties, preferentially water-wet rock.

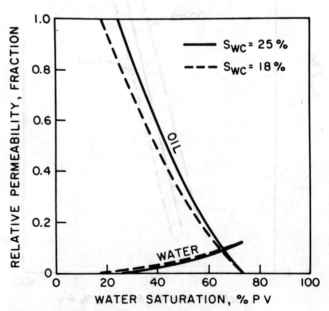

Fig. 2.21 Effect of irreducible connate water saturation on water-oil flow properties.

17. Johansen, R. T. and Dunning, H. N.: "Relative Wetting Tendencies of Crude Oil by the Capillarimetric Method", *Prod. Monthly* (Sept., 1959) **24,** No. 9.

18. Slobod, R. L. and Blum, H. A.: "Method for Determining Wettability of Reservoir Rocks", *Trans.*, AIME (1952) **195,** 1-4.

19. Brown, R. J. S. and Fatt, I.: "Measurements of Fractional Wettability of Oilfield Rocks by the Nuclear Magnetic Relaxation Method", *Trans.*, AIME (1956) **207,** 262-264.

20. Holbrook, O. C. and Bernard, G. G.: "Determination of Wettability by Dye Adsorption", *Trans.*, AIME (1958) **213,** 261-264.

21. Benner, F. C., Riches, W. W. and Bartell, F. E.: "Nature and Importance of Surface Forces in Production of Petroleum", *Drill. and Prod. Prac.*, API (1939).

22. Geffen, T. M., Owens, W. W., Parrish, D. R. and Morse, R. A.: "Experimental Investigation of Factors Affecting Laboratory Relative Permeability Measurements", *Trans.*, AIME (1951) **192,** 99-110.

23. Chatenever, A. and Calhoun, J. C., Jr.: "Visual Examinations of Fluid Behavior in Porous Media — Part I", *Trans.*, AIME (1952) **195,** 149-156.

24. "Fluid Distribution in Porous Systems — A Preview of the Motion Picture", Stanolind Oil and Gas Co. (1952); subsequently reprinted by Pan American Petroleum Corp. and Amoco Production Co.

25. Leverett, M. C.: "Capillary Behavior in Porous Solids", *Trans.*, AIME (1941) **142,** 159-172.

26. Welge, H. J.: "Displacement of Oil from Porous Media by Water and Gas", *Trans.*, AIME (1948) **179,** 133-138.

27. Purcell, W. R.: "Interpretation of Capillary Pressure Data", *Trans.*, AIME (1950) **189,** 369-374.

28. Jones-Parra, J.: "Comments on Capillary Equilibrium", *Trans.*, AIME (1953) **198,** 314-316.

29. Killins, C. R., Nielsen, R. F., and Calhoun, J. C., Jr.: "Capillary Desaturation and Imbibition in Rocks", *Prod. Monthly* (Feb., 1953) **18,** No. 2, 30-39.

30. Arman, I. H.: "Relative Permeability Studies", MS thesis, The U. of Oklahoma, Norman (1952).

31. Pickell, J. J., Swanson, B. F., and Hickman, W. B.: "Application of Air-Mercury and Oil-Air Capillary Pressure Data in the Study of Pore Structure and Fluid Distribution", *Soc. Pet. Eng. J.* (March, 1966) 55-61.

32. Brown, H. W.: "Capillary Pressure Investigations", *Trans.*, AIME (1951) **192,** 67-74.

33. Rose, W. R. and Bruce, W. A.: "Evaluation of Capillary Character in Petroleum Reservoir Rock", *Trans.*, AIME (1949) **186,** 127-133.

34. Richardson, J. G., Perkins, F. M., Jr., and Osoba, J. S.: "Differences in Behavior of Fresh and Aged East Texas Woodbine Cores", *Trans.*, AIME (1955) **204,** 86-91.

35. Mungan, N.: "Certain Wettability Effects in Laboratory Waterfloods", *J. Pet. Tech.* (Feb., 1966) 247-252.

36. Jennings, H. Y., Jr.: "Waterflood Behavior of High Viscosity Crudes in Preserved Soft and Unconsolidated Cores", *J. Pet. Tech* (Jan., 1966) 116-120.

37. Kyte, J. R., Naumann, V. O. and Mattax, C. C.: "Effect of Reservoir Environment on Water-Oil Displacements", *J. Pet. Tech.* (June, 1961) 579-582.

38. Johnston, N. and van Wingen, N.: "Recent Laboratory Investigation of Water Flooding in California", *Trans.*, AIME (1953) **198,** 219-224.

39. Buckley, S. E. and Leverett, M. C.: "Mechanism of Fluid Displacement in Sands", *Trans.*, AIME (1942) **146,** 107-116.

40. Hassler, G. L.: U. S. Patent No. 2,345,935 (April, 1944).

41. Morse, R. A., Terwilliger, P. L. and Yuster, S. T.: "Relative Permeability Measurements on Small Core Samples", *Oil and Gas J.* (Aug. 23, 1947); also Technical Paper 124, Pennsylvania State College Mineral Industries Experiment Station (1947).

42. Leas, W. J., Jenks, L. H. and Russell, C. D.: "Relative Permeability to Gas", *Trans.*, AIME (1950) **189,** 65-72.

43. Osoba, J. S., Richardson, J. G., Kerver, J. K., Hafford, J. A. and Blair, P. M.: "Laboratory Measurements of Relative Permeability", *Trans.*, AIME (1951) **192,** 47-56.

44. Hassler, G. L., Rice, R. R., and Leeman, E. H.: "Investigations on the Recovery of Oil from Sandstones by Gas Drive", *Trans.*, AIME (1936) **118,** 116-137.

45. Richardson, J. G., Kerver, J. K., Hafford, J. A., and Osoba, J. S.: "Laboratory Determination of Relative Permeability", *Trans.*, AIME (1952) **195,** 187-196.

46. Caudle, B. H., Slobod, R. L., and Brownscombe, E. R.: "Further Developments in the Laboratory Determination of Relative Permeability", *Trans.*, AIME (1951) **192,** 145-150.

47. Elkins, L. F.: Unpublished Communication (1943) cited by S. J. Pirson in *Elements of Oil Reservoir Engineering*, McGraw-Hill Book Co., Inc., New York (1950) 328.

48. Sandberg, C. R., Gournay, L. S. and Sippel, R. F.: "The Effect of Fluid-Flow Rate and Viscosity on Laboratory Determinations of Oil-Water Relative Permeabilities", *Trans.*, AIME (1958) **213,** 36-43.

49. Welge, H. J.: "A Simplified Method for Computing Oil Recovery by Gas or Water Drive", *Trans.*, AIME (1952) **195,** 91-98.

50. Stewart, C. R., Craig, F. F., Jr., and Morse, R. A.: "Determination of Limestone Performance Characteristics by Model Flow Tests", *Trans.*, AIME (1953) **198,** 93-102.

51. Owens, W. W., Parrish, D. R., and Lamoreaux, W. E.: "An Evaluation of a Gas Drive Method for Determining Relative Permeability Relationships", *Trans.*, AIME (1956) **207,** 275-280.

52. Johnson, E. F., Bossler, D. P., and Naumann, V. O.: "Calculation of Relative Permeability from Displacement Experiments", *Trans.*, AIME (1959) **216,** 370-372.
53. Kyte, J. R. and Rapoport, L. A.: "Linear Waterflood Behavior and End Effects in Water-Wet Porous Media", *Trans.*, AIME (1958) **213,** 423-426.
54. Kennedy, H. T. and Guerrero, E. T.: "The Effect of Surface and Interfacial Tensions on the Recovery of Oil by Water Flooding", *Trans.*, AIME (1954) **201,** 124-131.
55. Newcombe, J., McGhee, J. and Rzasa, M. J.: "Wettability Versus Displacement in Water Flooding in Unconsolidated Sand Columns", *Trans.*, AIME (1955) **204,** 227-232.
56. Mungan, N.: "Role of Wettability and Interfacial Tension in Waterflooding", *Soc. Pet. Eng. J.* (June, 1964) 115-123.
57. Wagner, O. R. and Leach, R. O.: "Effect of Interfacial Tension on Displacement Efficiency", *Soc. Pet. Eng. J.* (Dec., 1966) 335-344.
58. Fatt, I.: "The Effect of Overburden Pressure on Relative Permeability", *Trans.*, AIME (1953) **198,** 325-326.
59. McLatchie, A. S., Hemstock, R. A. and Young, J. W.: "The Effective Compressibility of Reservoir Rock and Its Effect on Permeability", *Trans.*, AIME (1958) **213,** 386-388.
60. Wyllie, M. R. J. and Gardner, G. H. F.: "The Generalized Kozeny-Carman Equation", *World Oil* (March and April, 1958) **146,** No. 4, 121-128 and No. 5, 210-228.
61. Corey, A. T., Rathjens, C. H., Henderson, J. H. and Wyllie, M. R. J.: "Three-Phase Relative Permeability", *Trans.*, AIME (1956) **207,** 349-351.
62. Land, C. S.: "Calculation of Imbibition Relative Permeability for Two- and Three-Phase Flow from Rock Properties", *Soc. Pet. Eng. J.* (June, 1968) 149-156.
63. Fatt, I. and Dykstra, H.: "Relative Permeability Studies", *Trans.*, AIME (1951) **192,** 249-256.
64. Fatt, I.: "The Network Model of Porous Media", *Trans.*, AIME (1956) **207,** 144-181.
65. Pirson, S. J., Boatman, E. M. and Nettle, R. L.: "Prediction of Relative Permeability Characteristics of Intergranular Reservoir Rocks from Electrical Resistivity Measurements", *J. Pet. Tech.* (May, 1964) 564-570.
66. Leverett, M. C. and Lewis, W. B.: "Steady Flow of Gas - Oil - Water Mixtures Through Unconsolidated Sands", *Trans.*, AIME (1941) **142,** 107-120.
67. Sarem, A. M.: "Three-Phase Relative Permeability Measurements by Unsteady-State Method", *Soc. Pet. Eng J.* (Sept., 1966) 199-205.
68. Saraf, D. N. and Fatt, I.: "Three-Phase Relative Permeability Measurement Using a Nuclear Magnetic Resonance Technique for Estimating Fluid Saturation", *Soc. Pet. Eng. J.* (Sept., 1967) 235-242.
69. Naar, J. and Wygal, R. J.: "Three-Phase Imbibition Relative Permeability", *Soc. Pet. Eng. J.* (Dec., 1961) 254-263.
70. Snell, R. W.: "Three-Phase Relative Permeability in an Unconsolidated Sand", *J. Institute of Petroleum* (March, 1962) **48,** 459.
71. Messer, E. S.: "Interstitial Water Determination by an Evaporation Method", *Trans.*, AIME (1951) **192,** 269-274.
72. Owens, W. W.: Private Communication.

Chapter 3

Efficiency of Oil Displacement by Water

In this chapter we shall deal with "oil displacement efficiency". The term refers to the portion of the oil initially in place that water sweeps from a unit volume of the reservoir. To discuss how water displaces oil from a reservoir of complex permeability and porosity, we must first understand the behavior of a reservoir segment having uniform properties.

3.1 Frontal Advance Theory

Fractional Flow Equation

In 1941, Leverett[1] in his pioneering paper presented the concept of fractional flow. Starting with the well known Darcy's law for water and for oil, he obtained

$$f_w = \frac{1 + \dfrac{k\,k_{ro}}{u_t\,\mu_o}\left(\dfrac{\partial P_c}{\partial L} - g\,\Delta\rho\,\sin\alpha_d\right)}{1 + \dfrac{\mu_w}{\mu_o}\dfrac{k_o}{k_w}}, \quad (3.1)$$

where

- f_w = fraction of water in the flowing stream passing any point in the rock (i.e. the watercut)
- k = formation permeability
- k_{ro} = relative permeability to oil
- k_o = effective permeability to oil
- k_w = effective permeability to water
- μ_o = oil viscosity
- μ_w = water viscosity
- u_t = total fluid velocity (i.e., q_t/A)
- P_c = capillary pressure = $p_o - p_w$ = pressure in oil phase minus pressure in water phase
- L = distance along direction of movement
- g = acceleration due to gravity
- $\Delta\rho$ = water-oil density differences = $\rho_w - \rho_o$
- α_d = angle of the formation dip to the horizontal

A derivation of this equation is presented in Appendix A. (Eq. 3.1 in this chapter is identical with Eq. A.11 in Appendix A.) The terms in this equation are in consistent units; i.e., darcy, cp, cm/sec, cm, atm/cm, cm/sq sec, gm/cc. Note that some of the signs in Eq. 3.1 are slightly different from their counterparts in the equation derived by Leverett. This is caused by the variation in definitions of capillary pressure and water-oil density difference.

In so-called practical units, Eq. 3.1 becomes

$$f_w = \frac{1 + 0.001127\,\dfrac{k\,k_{ro}}{\mu_o}\dfrac{A}{q_t}\left[\dfrac{\partial P_c}{\partial L} - 0.433\,\Delta\rho\,\sin\alpha_d\right]}{1 + \dfrac{\mu_w}{\mu_o}\dfrac{k_o}{k_w}}$$

$$\ldots\ldots\ldots\ldots (3.1a)$$

where permeability is in md; viscosity, cp; area, sq ft; flow rate, B/D; pressure, psi; distance, ft; and density difference, gm/cc.

Note that the fractional flow of water, f_w, for a given set of rock, formation and flooding conditions is a function of water saturation alone. This arises because the relative permeability and capillary pressure characteristics are functions of saturation alone.

All the factors necessary to calculate the value of f_w are readily available except one, the capillary pressure gradient. This gradient can be expressed as

$$\frac{\partial P_c}{\partial L} = \frac{\partial P_c}{\partial S_w}\frac{\partial S_w}{\partial L} \quad\ldots\ldots\ldots (3.2)$$

Although the value of $\dfrac{\partial P_c}{\partial S_w}$ can be determined from the appropriate water-oil capillary pressure curve, the value of the saturation gradient, $\dfrac{\partial S_w}{\partial L}$, is not available; so in practical use the capillary pressure term in Eq. 3.1 is neglected (but not forgotten). Then Eq. 3.1 simplifies to

$$f_w = \frac{1 - \frac{k}{u}\frac{k_{ro}}{\mu_o}(g \triangle \rho \sin \alpha_d)}{1 + \frac{\mu_w}{\mu_o}\frac{k_o}{k_w}} \quad \ldots \quad (3.3)$$

For the further simplification where displacement occurs in a horizontal system, this equation reduces to

$$f_w = \frac{1}{1 + \frac{\mu_w}{\mu_o}\frac{k_o}{k_w}} \quad \ldots \ldots \ldots \quad (3.4)$$

This equation, which is termed the simplified form of the fractional flow equation, is identical with Eq. A.12 in Appendix A. It involves the water and oil relative permeabilities as a ratio. Therefore, this equation can also be shown as

$$f_w = \frac{1}{1 + \frac{\mu_w}{\mu_o}\frac{k_{ro}}{k_{rw}}}, \quad \ldots \ldots \quad (3.4a)$$

where k_{ro} and k_{rw} are the relative permeabilities to oil and water, respectively.

The term f_w is a function of water saturation inasmuch as both k_{ro} and k_{rw} are themselves functions of water saturation. At increasing water saturations, the value of k_{ro} declines, whereas that of k_{rw} rises, with the result that the value of f_w increases.

Eq. 3.4a can also be obtained more simply. By definition, f_w is the water flow rate divided by the total flow rate, or

$$f_w = \frac{q_w}{q_w + q_o}, \quad \ldots \ldots \ldots \quad (3.5)$$

where q_o and q_w are the oil and water flow rates, respectively.

If we divide both the numerator and denominator by q_w, we obtain

$$f_w = \frac{1}{1 + \frac{q_o}{q_w}}. \quad \ldots \ldots \ldots \quad (3.6)$$

However, since the oil-water ratio q_o/q_w can be expressed in terms of the water and oil relative permeabilities as

$$\frac{q_o}{q_w} = \frac{k_{ro}}{\mu_o}\frac{\mu_w}{k_{rw}}, \quad \ldots \ldots \ldots \quad (3.7)$$

substituting Eq. 3.7 into Eq. 3.6 gives Eq. 3.4a.

Eqs. 3.4 and 3.4a, both of which apply to a horizontal system, indicate that for a given rock—that is, a given set of water-oil relative permeability characteristics—the value of f_w depends upon the magnitude of oil and water viscosities. (This effect is discussed in Section 3.7.)

The relative permeability data shown in Figs. 2.16 and 2.17 apply for water-wet and oil-wet systems, respectively. Using both these basic flow properties, the values of fractional flow have been calculated for an oil-water viscosity ratio of 2. These values are shown in Figs. 3.1 and 3.2 for both strongly water-wet and

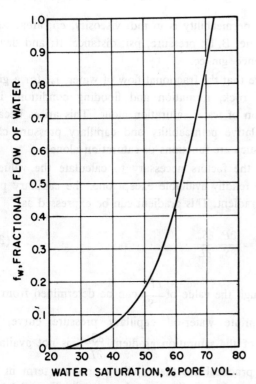

Fig. 3.1 Fractional flow curve, strongly water-wet rock. $\mu_o = 1$ cp; $\mu_w = 0.5$ cp

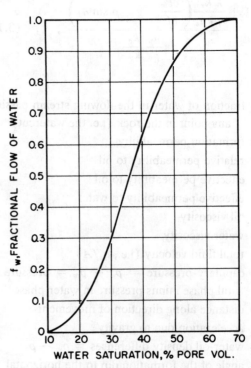

Fig. 3.2 Fractional flow curve, strongly oil-wet rock. $\mu_o = 1$ cp; $\mu_w = 0.5$ cp

strongly oil-wet systems. At this oil-water viscosity ratio, the fractional flow curve for the strongly water-wet rock is concave upward; that for the oil-wet rock is concave upward at low water saturations and concave downward at the higher saturations. However, as we shall see in Section 3.7, the fractional flow curve for a preferentially water-wet rock becomes more similar to that for an oil-wet rock as the oil viscosity increases.

For inclined reservoirs (Eq. 3.3) the fractional flow curve is in addition dependent upon the formation permeability, the total flow rate, the density difference and the angle of dip. The sign of the water-oil density difference as defined is positive. The value of the angle of dip, α_d, is measured from the horizontal, with flow moving updip assigned a positive angle and flow moving downdip having a negative angle. Thus Eq. 3.3 shows that water displacing oil updip will result in a lower value of f_w at any water saturation than will water displacing oil downdip.

The effect of dip on the fractional flow curve is shown in Figs. 3.3 and 3.4. As before, these figures pertain to preferentially water-wet and oil-wet systems, respectively.

When calculating the fractional flow of water displacing oil updip, it is possible to obtain negative values for f_w. For Eq. 3.3 it is assumed, of course, that the capillary pressure gradient is negligible. The physical significance of a negative value of f_w is that in the absence of a capillary pressure gradient the tendency of water existing at these saturations is to flow *downward*—that is, gravity forces segregate the oil and water. However, as we shall see, when capillary forces are included these calculated negative values of f_w have no effect upon the displacement process.

Let us now turn our attention to the qualitative effect of the capillary pressure gradient upon the fractional flow. Separating the capillary pressure gradient into its two components—the change of capillary pressure with water saturation and the saturation gradient with length (Eq. 3.2)—we must observe that the saturation gradient with length is always negative. That is, the water saturation decreases with distance in the direction of flow. Referring to Figs. 2.9 through 2.12, note that the capillary pressure change with water saturation for conditions in which the water saturation is increasing is always negative. That is, the capillary pressure decreases with increasing water saturation. Thus the capillary pressure gradient, dP_c/dL, has a positive sign and its effect is to increase the value of the fractional flow curve in the saturation range where capillary effects are significant. Also from Eq. 3.2, it can be deduced that where the saturation gradient is low, the capillary pressure gradient will also be low.

Frontal Advance Equation

The 1942 paper[2] of Buckley and Leverett presented the frontal drive equation. Considering a small element within a continuous porous medium, they expressed the

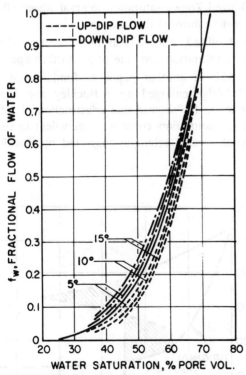

Fig. 3.3 Effect of formation dip on fractional flow curve, strongly water-wet rock.
$\mu_o = 1$ cp; $\mu_w = 0.5$ cp; $k = 400$ md;
$u_t = 0.01$ B/D/sq ft

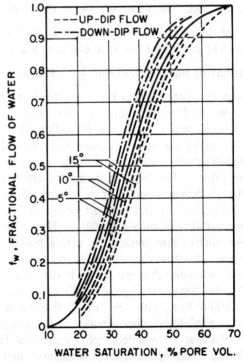

Fig. 3.4 Effect of formation dip on fractional flow curve, strongly oil-wet rock.
$\mu_o = 1$ cp; $\mu_w = 0.5$ cp; $k = 400$ md;
$u_t = 0.01$ B/D/sq ft

difference at which the displacing fluid enters this element and the rate at which it leaves it in terms of the accumulation of displacing fluid. By transforming this material balance equation, the frontal advance equation can be obtained:

$$\left(\frac{\partial L}{\partial t}\right)_{S_w} = \frac{q_t}{A\phi}\left(\frac{\partial f_w}{\partial S_w}\right)_t \quad \ldots \ldots \quad (3.8)$$

A derivation of this equation is presented in Appendix B. The only assumptions necessary for the derivation are that (1) there is no mass transfer between phases, and (2) the phases are incompressible.

This equation states that the rate of advance of a plane of fixed water saturation is equal to the total fluid velocity multiplied by the change in composition of the flowing stream caused by a small change in the saturation of the displacing fluid. That is, any water saturation, S_w, moves along the flow path at a velocity equal to $\frac{q_t}{A\phi}\frac{df_w}{dS_w}$. As the total flow rate q_t increases, the velocity of the plane of saturation correspondingly increases. As the total flow rate is reduced, the velocity of the saturation is correspondingly reduced.

Eq. 3.8 can be integrated to yield

$$L = \frac{W_i}{A\phi}\left(\frac{df_w}{dS_w}\right), \quad \ldots \ldots \ldots \quad (3.9)$$

where L is the total distance that the plane of given water saturation moves.

Buckley and Leverett pointed out[2] that Eq. 3.9 can be used to calculate the saturation distribution existing during a flood. The value of $\frac{df_w}{dS_w}$ is the slope of the curve of fractional flow vs water saturation. Most fractional flow curves exhibit two water saturations having the same value of $\frac{df_w}{dS_w}$. The consequence of this is that according to Eq. 3.8, two different saturations would have the same velocity—that is, would exist at the same point in the formation at the same time. To make the situation even more absurd, if an initial saturation gradient exists above the water-oil contact prior to flooding, the computed saturation distribution is triple-valued over a portion of its length (Fig. 3.5). Buckley and Leverett recognized the physical impossibility of such a condition. They pointed out that the correct interpretation is that a portion of the computed saturation distribution is imaginary and that the real saturation-distance curve is discontinuous. In Fig. 3.5 the "imaginary" portion of the curve is shown as a dashed line, with the "real" distribution curve shown as a solid line, discontinuous at L_1. The position of the plane at L_1 is determined by material balance, the shaded areas between the "imaginary" and "real" curves to the left and right of L_1 being equal. Buckley and Leverett recognized that the capillary pressure gradient, neglected in their use of the fractional flow curve, would become exceedingly large at the saturation discontinuity. As a result the plane of saturation discontinuity would be converted into a zone of more gradual transition in saturation, the width of this zone depending upon the flow rate.

Because of the "triple-valued curve" and its significance, some investigators were hesitant to use the Buckley-Leverett frontal advance equation.

Finally, in 1951 two papers[3,4] were published that made use of the frontal advance equation. In the second of these papers Terwilliger et al.[4], in a paper of fundamental significance, studied the application of the fractional flow and frontal advance equations to gas-oil gravity drainage performance. They found that at the lower range of displacing (gas) saturations, these saturations all moved downward at the same velocity, with the result that the shape of the saturation distribution over this range of saturations was constant with time. They termed this saturation distribution the "stabilized zone". They found that using the complete form of the fractional flow equation (Eq. 3.1), together with the steady-state gas-oil relative permeability characteristics and capillary pressure characteristics, agreement was obtained between the calculated and the observed saturation distributions. In addition the authors showed that by laying a tangent to the fractional flow curve from S_w corresponding to the initial displacing fluid saturation and f_w equal to zero they could define the saturation at the upstream end of the stabilized zone. As a result of this paper, there developed the concept of the stabilized and nonstabilized zones, defined as follows:

Stabilized Zone—saturation interval where all points of saturation move at the same rate.

Nonstabilized Zone—saturation interval where all points of saturation continue to get farther apart.

In 1952 in another paper of fundamental importance,[5] Welge enlarged upon Buckley and Leverett's earlier work. He showed that construction of a tangent to the fractional flow curve was equivalent to the "balancing of areas" technique suggested by Buckley and

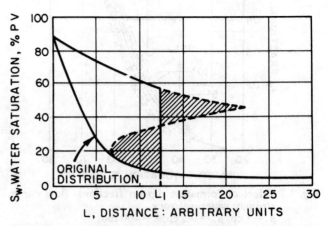

Fig. 3.5 Triple-valued saturation distribution (after Ref. 2).

Leverett[2] for finding the saturation at the "discontinuity". Welge further went on to derive an equation that relates the average displacing fluid saturation to that saturation at the producing end of the system. Thus, in waterflooding terminology,

$$\bar{S}_w - S_{w2} = Q_i f_{o2}, \quad \ldots \ldots \quad (3.10)$$

where

$\bar{S}_w$ = average water saturation, fraction PV

S_{w2} = water saturation at the producing end of the system, fraction PV

Q_i = pore volumes of cumulative injected fluid, dimensionless

f_{o2} = fraction of oil flowing at the outflow end of the system.

(See Appendix C for an alternative derivation of Eq. 3.10.) This equation is important because it relates three factors of prime interest in waterflooding: (1) the average water saturation and hence the total oil recovery, (2) the cumulative injected water volume, and (3) the oil cut and thus the water cut and producing WOR.

Welge[5] also determined that

$$Q_i = \frac{1}{\left(\dfrac{df_w}{dS_w}\right)_{S_{w2}}} \quad \ldots \ldots \quad (3.11)$$

This corollary equation makes it possible to relate independently the cumulative water injected and the producing end water saturation.

In 1952 Kern[6] transformed to radial coordinates the Buckley-Leverett equation developed originally for a linear system. He also suggested a technique for eliminating the graphical integration required by Buckley and Leverett's "balancing of areas" techniques.

(We shall defer any discussions on the practical application of the fractional flow and frontal advance equation until later in this section.)

Mathematical Studies

A few years later a series of mathematical studies[7,8] dealt with the theory of two-phase flow and displacement in porous media. With the advent of high-speed computers, numerical studies of oil displacement including the effects of capillary pressure became possible.[9,10]

Interest appeared to be concentrated on the significance of Buckley-Leverett's triple-valued saturation.[11] Later work[12,13] confirmed that the inclusion of capillary pressure effects, particularly at the flood front, was all that was necessary to eliminate the bothersome triple value.

Douglas et al.[14] showed a comparison of the saturation distribution calculated by the Buckley-Leverett frontal advance equation with that determined from computer calculations including capillary effects. Later it was confirmed[15] that numerical calculations yielded the same results as the Buckley-Leverett theory.

This limited discussion on the mathematical and numerical studies does not suggest that their value is limited. These mathematical studies, together with the experimental studies to be discussed next, increase the confidence of reservoir engineers in the general usefulness of the frontal drive equation.

Experimental Confirmation

In 1954, Levine[16] reported the results of some detailed experimental studies using a large alundum core. In these experiments, the pressure in the oil phase was measured independent of that in the water phase, so that the pressure distribution in each phase was determined. His experimental apparatus also permitted the measurement of the saturation distribution along the core. Levine confirmed both that relative permeability was not dependent upon the fluid viscosities and that it was affected by the direction of the saturation change. The use of the fractional flow and frontal advance equations gave a reasonable agreement between the calculated and experimentally determined oil recoveries at water breakthrough. To get this agreement, though, it was necessary to include capillary pressure and gravity terms in the fractional flow equation.

In 1956, Owens et al.[17] compared the gas-oil relative permeability characteristics measured by the steady-state technique with those determined from the performance of an external gas drive.[5] The agreement between the flow properties measured in two different ways confirms the validity of the frontal advance equation upon which is based the external drive method of determining relative permeability characteristics. A similar comparison[18] of water-oil relative permeability characteristics of preferentially water-wet rocks is also an implied proof. In another study[19] it was demonstrated that the frontal advance equation could be used to accurately predict waterflood performance.

Stabilized Zone Effects

As a result of the theoretical and experimental analyses of the displacement of oil by an injected fluid, either water or gas, it was generally recognized that two distinct bands of saturation move through the reservoir. The first is a zone or bank of high saturation gradient that exists at low displacing fluid saturations. It has been termed the "stabilized zone" or "primary phase". This is followed by a gradual oil displacement—that is, a region with much lower saturation gradient—sometimes termed the "subordinate" phase of the flood. It was recognized that the stabilized zone represented some form of dynamic balance between capillary and viscous effects. In the early 1950's some conflicting evidence arose on the significance of the length of the stabilized zone.

Rapoport and Leas[20] set out to make a comprehensive investigation of the stabilized zone effect. They found that because of the stabilized zone, the performance of a laboratory waterflood depends on both the injection rate and the length of the system. From both theoretical and experimental considerations they found that as the term $Lu\mu_w$ increases in value, the flooding behavior becomes independent of rate and length, and thus is "stabilized". It is to these stabilized conditions that the theory formulated by Buckley and Leverett is applicable. Rapoport and Leas further conclude that under field conditions the flooding behavior is usually stabilized.

Jones-Parra and Calhoun,[21] in extending the work of Rapoport and Leas, suggested that the approach to stabilization of a waterflood depends upon the term $Lu\mu_w/(\sqrt{k\phi}\ \sigma \cos \theta)$. Bail[22] also showed that at conventional flooding rates, the length of the stabilized zone becomes small and is of negligible effect. Also in that paper, he presented a detailed outline of a method of calculating the length of the stabilized zone for any flooding conditions.

3.2 Water Tonguing

In 1953, Dietz[23] proposed what might be termed a competitor to the Buckley-Leverett frontal advance theory. He was particularly concerned with edge-water encroachment but pointed out that this theory could also be applied to substantially horizontal systems. Dietz visualized that water displaces oil by underrunning the oil, or "tonguing". He visualized in the water-invaded region a residual oil saturation with only water moving, and in the uninvaded region, only oil flowing. There would be a sharp interface of negligible extent separating the flowing water from the flowing oil (Fig. 3.6).

In 1955, the results of a series of experimental waterfloods[24] permitted evaluation of the waterflooding theories of Buckley-Leverett and Dietz. Although the Dietz theory was found to be in agreement with some of the experimental results, the Buckley-Leverett theory was successful in correlating all the experimental results.

3.3 Viscous Fingering

The Buckley-Leverett frontal advance theory assumes

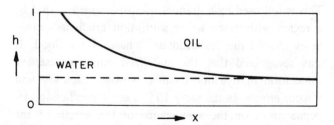

Fig. 3.6 Shape of the water tongue according to Dietz.[23]

that the initial displacement of oil by water occurs as a smooth, substantially straight interface. In 1951, Engelberts and Klinkenberg[25] showed that in scaled experiments the existence of fingers, or discrete streamers of displacing water moving through the oil, could be inferred. Later van Meurs[26] demonstrated the existence of these fingers by a clever experimental technique (Fig. 3.7). He showed that even in laboratory systems where care is taken to insure as nearly uniform porous media as possible, the tendency for these fingers to form increases as the oil-water viscosity ratio increases.[27] Later work by a number of other investigators[28-31] showed that at high oil-water viscosity ratios, instabilities exist at the oil-water interface and these instabilities grow until their effect is dominant on the over-all flooding performance.

Debates raged over the practical significance of these fingers. An otherwise peaceful gathering of production research engineers could be turned into a heated controversy by the question, "Do you think viscous fingers are solely a laboratory phenomenon?" It is my opinion that in the actual reservoir having complex variations in permeability and porosity a macroscopically nonuniform flood front very much similar to viscous fingering will occur, if from no other cause than the nonuniformities in rock permeabilities. A practicing reservoir engineer would term this the effect of reservoir heterogeneities (see Chapter 6 and following).

3.4 Mobility of the Connate Water

In waterflooding operations there was some question as to the extent to which the reservoir connate water is contacted and displaced by invading flood water. In 1957 Brown[32] reported on a study in preferentially water-wet rocks. He found that it is the connate water in the reservoir that actually displaces the oil, the connate water itself being displaced by the injected water. This finding was confirmed[33] for preferentially water-wet rocks.

Unfortunately, results of similar studies for preferentially oil-wet rocks are not available. However, it is probable that since the connate water in oil-wet rocks does not affect the water-oil relative permeability characteristics, the injected water does not contact the connate water to any significant degree.

A ramification of this concern about the mobility of connate water is that reservoir wettability can be determined from field performance. If the water produced upon water breakthrough at the producing wells has a mineral composition the same as that of the connate water, the reservoir is likely to be water-wet. If the water produced at water breakthrough is injected water, two possibilities are likely: (1) the reservoir is preferentially oil-wet, and (2) the injected water is moving through very thin porous zones or fractures and, because of this, has no opportunity to contact significant quantities of connate water. Of course, the second pos-

EFFICIENCY OF OIL DISPLACEMENT BY WATER

sibility can be eliminated if water breakthrough does not occur early. It is possible that in water-wet reservoirs of unique heterogeneity the connate water displaced in the permeable layers imbibes into the tighter layers with oil counterflow back into the permeable layer. With this situation, injected water could be produced at or shortly after water breakthrough.

3.5 Practical Application of the Frontal Advance Theory

The information necessary to apply the Buckley-Leverett frontal advance theory consists of only the appropriate water-oil relative permeability characteristics and the oil and water viscosities. The first step is to calculate the fractional flow curve, including the effect of formation dip if necessary. For a horizontal system, Eq. 3.4a is used; if formation dip effects are to be included, Eq. 3.3 is used.

$$f_w = \frac{1}{1 + \frac{\mu_w}{\mu_o} \frac{k_{ro}}{k_{rw}}}, \quad \quad (3.4a)$$

$$f_w = \frac{1 - \frac{k}{u_t} \frac{k_{ro}}{\mu_o}(g\,\Delta\rho\,\sin\alpha_d)}{1 + \frac{\mu_w}{\mu_o} \frac{k_{ro}}{k_{rw}}} \quad \quad (3.3)$$

The fractional flow curve is then differentiated to obtain either a curve or tabulated values of df_w/dS_w vs water saturation, S_w.

To illustrate the use of the fractional flow curve we shall use one of those shown originally as Fig. 3.2, which applies to an oil-wet system. Fig. 3.8 shows a replot of the fractional flow curve shown in Fig. 3.2. Fig. 3.9 shows the corresponding df_w/dS_w curve. Note that the slope of the fractional flow curve is finite at the maximum water saturation. This indicates that the maximum water saturation advances at a finite velocity.

Recovery at Water Breakthrough

For an initially liquid-saturated, linear system of length L at the time of water breakthrough to the producing end, Eq. 3.9 can be written as

$$L = \frac{W_i}{A\phi}\left(\frac{df_w}{dS_w}\right)_f, \quad \quad (3.12)$$

where the subscript f denotes the condition at flood front. The total oil displaced is equal to W_i. Therefore, the average water saturation at breakthrough is the sum of the connate water saturation and the increase in water saturation caused by waterflooding, or

$$\bar{S}_{wbt} = S_{wc} + \frac{W_i}{A\phi L} \quad \quad (3.13)$$

Substituting Eq. 3.12 into Eq. 3.13 and transforming,

$$\bar{S}_{wbt} - S_{wc} = \frac{1}{\left(\dfrac{df_w}{dS_w}\right)_f} = \frac{S_{wf} - S_{wc}}{f_{wf}}. \quad (3.14)$$

$N_p = W_i = 6.0\%$

$N_p = 20\%;\ W_i = 34\%$

$N_p = W_i = 12\%$

$N_p = 52\%;\ W_i = 650\%$

Fig. 3.7 Viscous fingering.[26]

This equation indicates that laying a tangent to the fractional flow curve from S_w corresponding to the connate water saturation and f_w equal to zero yields at the point of tangency the water saturation at the front, S_{wf}. Extrapolating this tangent to $f_w = 1.0$ yields the value of the average water saturation at breakthrough (Fig. 3.10).

Performance After Water Breakthrough

Upon continued water injection after breakthrough, the equations developed by Welge[5] are of value. They are:

$$\bar{S}_w - S_{w2} = Q_i f_{o2}, \quad \ldots \ldots \quad (3.10)$$

and

$$Q_i = \frac{1}{\left(\dfrac{df_w}{dS_w}\right)_{S_{w2}}} \quad \ldots \ldots \quad (3.11)$$

In using these equations we find that a tangent to the fractional flow curve has the following properties:

1. The tangent point, S_{w2}, represents the water saturation at the producing end of the system.

2. The value of fractional flow, f_w, at the point of tangency is the producing water cut.

3. The saturation at which the tangent intersects $f_w = 1.0$ is the average water saturation.

4. The inverse of the slope of the tangent is equal to

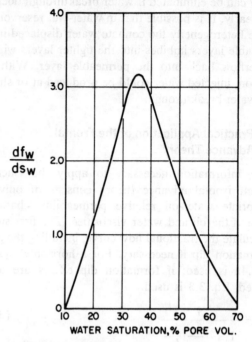

Fig. 3.9 Plot of df_w/dS_w for oil-wet rock. $\mu_o = 1$ cp; $\mu_w = 0.5$ cp

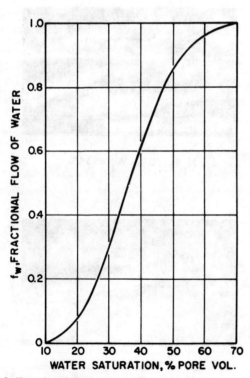

Fig. 3.8 Fractional flow curve, oil-wet rock. $\mu_o = 1$ cp; $\mu_w = 0.5$ cp

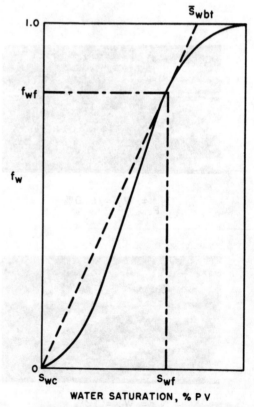

Fig. 3.10 Determination of average water saturation at breakthrough, $\bar{S}_{wbt}$.

EFFICIENCY OF OIL DISPLACEMENT BY WATER

the cumulative injected fluid in pore volumes; i.e., Q_i.

Although the waterflood performance after breakthrough can be demonstrated using the fractional flow curve and tangent construction, this is generally done more accurately using Eqs. 3.10 and 3.11. Fig. 3.11 shows the calculated waterflood performance using these equations.

Saturation Distribution During Waterflooding

It is sometimes of interest to determine the saturation gradient during a waterflood of a system having uniform permeability. To do this, Eq. 3.9 is used.

$$L = \frac{W_i}{A\phi}\left(\frac{df_w}{dS_w}\right) \qquad (3.9)$$

This equation indicates that at the time a volume of water W_i has been injected, the distance any saturation has moved depends upon the value of both W_i and df_w/dS_w at that saturation.

Fig. 3.12 shows a typical saturation distribution calculated at different cumulative injected water volumes using Eq. 3.9. Since the maximum water saturation has a finite value of df_w/dS_w, it then has a finite velocity. Each lower water saturation will have a higher velocity, and thus will have moved a greater distance. The water saturation having the highest velocity—that is, the highest value of slope—is that at the front, S_{wf}. Fig. 3.12 also shows a short stabilized zone ahead of the flood front.

The water saturation gradient is frequently concave up, sometimes nearly constant (as shown in Fig. 3.12) and only rarely concave down. The shape of the saturation distribution curve is related through Eq. 3.9 to the shape of the df_w/dS_w-water saturation curve (Fig. 3.9).

Modification for the Presence of a Mobile Connate Water Saturation

When the connate water is mobile, water will flow in the reservoir and be produced prior to water injection. The original producing water cut will correspond to the value of f_w from the fractional flow curve at the mobile connate water saturation. Let us call this value f_{w1}. Upon injection of water, the original water cut will continue until the waterflood front appears at the production well. As before, for a linear system of length L,

$$L = \frac{W_i}{A\phi}\left(\frac{df_w}{dS_w}\right)_f \qquad (3.12)$$

The total water injected, W_i, will result in an increase in water saturation of $\frac{W_i}{A\phi L}(1 - f_{w1})$. Thus

$$\overline{S}_{wbt} = S_{wc} + \frac{W_i}{A\phi L}(1 - f_{w1}) \qquad (3.15)$$

Substituting Eq. 3.12 into Eq. 3.15,

$$\overline{S}_{wbt} - S_{wc} = \frac{1 - f_{w1}}{\left(\dfrac{df_w}{dS_w}\right)_f} \qquad (3.16)$$

and

$$\left(\frac{df_w}{dS_w}\right)_f = \frac{1 - f_{w1}}{\overline{S}_{wbt} - S_{wc}} \qquad (3.16a)$$

This equation denotes that a tangent to the fractional flow curve from the value of the connate water saturation (S_{wc}) and the corresponding value of fractional flow of water (f_{w1}) will intersect $f_w = 1.0$ at the average water saturation at breakthrough of the flood front. The water saturation at this front is identical with that at the point of tangency (Fig. 3.13).

We can see from Fig. 3.13 that at increased values of connate water saturation there will be some value above which a tangent cannot be drawn to the fractional flow curve. That is, there is a maximum mobile

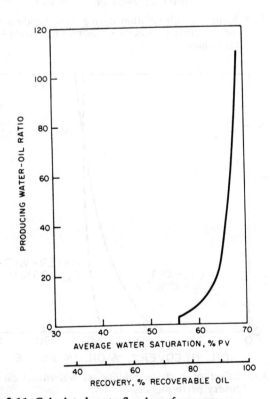

Fig. 3.11 Calculated waterflood performance.

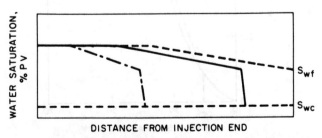

Fig. 3.12 Saturation distribution during waterflooding with increasing injected water volumes.

connate water saturation above which a waterflood front will not form. At this "critical" saturation, the connate water and the oil have equal mobilities; at connate water saturations below this value the oil is more mobile than the connate water and above this value the connate water is more mobile than the oil. It can be shown that at this saturation the oil and water mobilities are equal; that is, the value of f_w is 0.5. If we cannot draw a tangent, then we know that a flood front will not form.

Fig. 3.14 shows the saturation distributions with different injected water volumes for two systems in which there is mobile water saturation, the first where a flood front is formed and the second where no flood front exists.

3.6 Influence of Rock Wettability on Oil Production Performance

We are now in a position to see the influence of rock wettability on oil recovery performance. Using the typical water-wet and oil-wet flow properties shown in Figs. 2.16 and 2.17, respectively, we can calculate the waterflood performance of a linear system as shown in Fig. 3.15. Oil and water viscosities of 5 cp and 0.5 cp, respectively, were used in these calculations. Note that in a preferentially water-wet system the oil is recovered at a lower WOR and consequently with less injected water than in an oil-wet system. That is, the displacement of a wetting fluid by a nonwetting one is less efficient than the displacement of a nonwetting by a wetting fluid, all other things being equal.[34]

3.7 Influence of Oil and Water Viscosities

For an illustration of the effect of oil and water viscosities on the waterflood performance of different wettability systems, refer to Figs. 3.16 and 3.17. These figures demonstrate that regardless of the system's wettability, a higher oil viscosity results in less efficient displacement; that is, there is a lower recovery at any WOR, and an increased injected water volume is required to achieve that recovery.

Although the fractional flow curves in Figs. 3.16 and

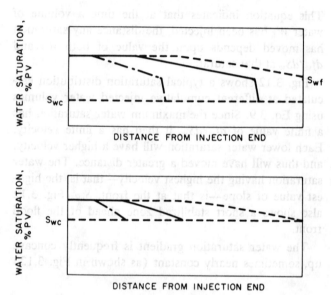

Fig. 3.14 Saturation distribution during waterflooding with mobile connate water with increasing injected water volumes.

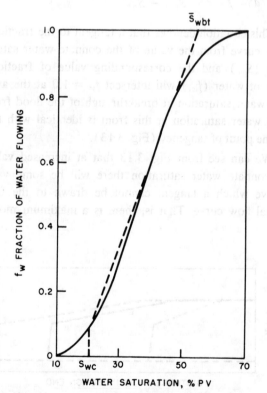

Fig. 3.13 Average water saturation at water breakthrough with mobile connate water present.

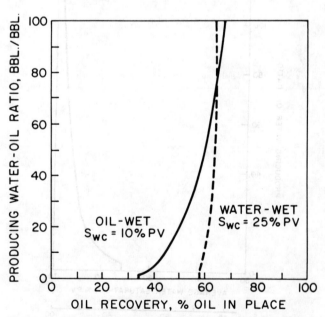

Fig. 3.15 Effect of rock wettability on waterflood oil recovery performance.
Oil viscosity = 5 cp; water viscosity = 0.5 cp.

EFFICIENCY OF OIL DISPLACEMENT BY WATER

3.17 show that the water-wet curves are generally concave up, and those for the oil-wet rock are concave down, this is not a correct generalization. It is my experience that the fractional flow curves for oil-wet rocks are, indeed, always concave down; but it is also my experience that the curves for preferentially water-wet rocks are frequently concave down, too.

3.8 Influence of Formation Dip and Rate

Referring to Eq. 3.1, we find that formation dip and rate have an interrelated effect on the fractional flow curve. The effect of rate and dip are shown in Figs. 3.18 and 3.19. These correspond to the different wettability systems and are based on the fractional flow curves shown in Figs. 3.1 and 3.2.

We observe that when water displaces oil updip, a more efficient performance is obtained at the lower flow rates; that is, the gravity force dominates. However, when the oil is displaced downdip, higher rates give the improved performance, since there is less tendency for water to percolate down, by gravity, through the oil.

Dip exerts another influence that we can observe: as the formation dip increases at any flow rate, the performance of a waterflood with oil being displaced updip improves, but with an oil being displaced downdip, the performance efficiency decreases.

3.9 Influence of Initial Gas Saturation

In a reservoir partially depleted by solution gas drive, an initial gas saturation will exist before waterflooding. Early investigators[2,3,35-37] found that the waterflooding of a linear system results in the formation of an oil bank, or zone of increased oil saturation ahead of the injected water. The moving oil bank will displace a portion of the free gas ahead of it, trapping the rest. The movement of the residual oil will cause the pressure on this trapped gas to be increased above the pressure at which it was trapped. The amount that the pressure on the gas will be increased before the waterflood front reaches it will depend upon the width of the oil bank, the oil permeability, and the flow rate. As the pressure increases, the gas is reduced in volume because of both compression and solution. It was also found that the effect of an initial gas saturation on oil displacement depends upon the magnitude of the trapped gas existing at the waterflood front. If no trapped gas remains in an element of the rock as the waterflood front reaches it, there is no influence on oil displacement.

Fig. 3.20 shows a saturation profile during waterflooding of a partially depleted reservoir. This sketch shows that the gas saturation at the leading edge of the oil bank is sharply reduced as a portion of the initial free gas is trapped. This gas is subsequently dissolved in the oil as it experiences the increasing pressure due to flooding.

Reduction of a Trapped Gas Saturation by Compression and Solution Within the Oil Bank

Let us consider that the gas saturation trapped at the

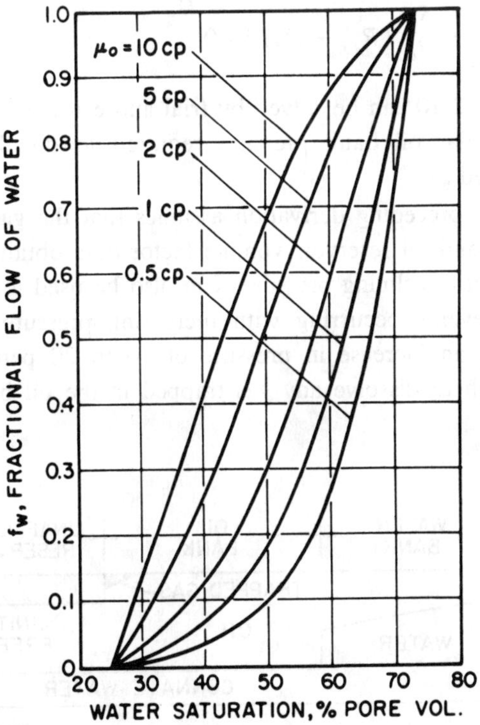

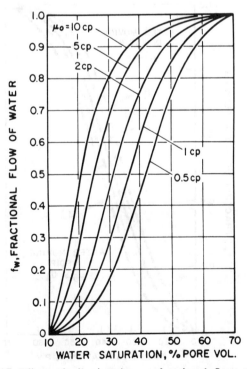

Fig. 3.16 Effect of oil viscosity on fractional flow curve, strongly water-wet rock.

Fig. 3.17 Effect of oil viscosity on fractional flow curve, strongly oil-wet rock.

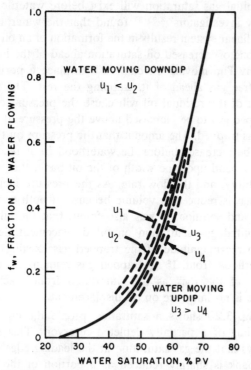

Fig. 3.18 Effect of rate on fractional flow curve, water-wet rock, dipping reservoir.

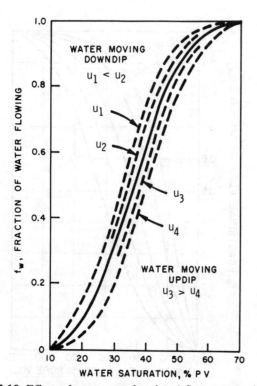

Fig. 3.19 Effect of rate upon fractional flow curve, oil-wet rock, dipping reservoir.

leading edge of the oil bank is S_{gt}, with a corresponding oil saturation, S_o, and a connate water saturation, S_{wc}. The gas present in a unit of pore space expressed at standard conditions is $(S_{gt}/B_g) + (S_o/B_o) R_s$. Let us consider that a volume of additional oil, $\triangle S_o$, saturated with gas at an increased pressure enters the unit volume, increasing the pressure to that same level. As a consequence, a portion of the trapped gas is compressed and dissolved. The resulting oil saturation is S_o', the oil having a formation volume factor of B_o' and a solution GOR of R_s'. The trapped gas, now at a higher pressure, S_{gt}', has a formation volume factor of B_g'. A volume balance yields

$$S_o \frac{B_o'}{B_o} + \triangle S_o + S_{gt}' + S_{wc} = 1 \quad \ldots \quad (3.17)$$

A gas balance in this unit volume, expressed at standard conditions, yields

$$\frac{S_{gt}}{B_g} + \frac{S_o}{B_o} R_s + \frac{\triangle S_o}{B_o'} R_s' = \frac{S_{gt}'}{B_g'} + \frac{S_o'}{B_o'} R_s'.$$
$$\ldots \ldots \ldots \ldots \ldots \ldots (3.18)$$

Combining Eqs. 3.17 and 3.18 to eliminate the term $\triangle S_o$ and to solve for S_{gt}' yields

$$S_{gt}' = B_g' \left[\frac{S_{gt}}{B_g} - \frac{S_o}{B_o} (R_s' - R_s) \right] \quad \ldots \quad (3.19)$$

Eq. 3.19 can be solved to determine the pressure level at which the trapped gas saturation is completely dissolved; i.e. $S_{gt}' = 0$. Thus

$$\frac{S_{gt}}{B_g} - \frac{S_o}{B_o} (R_s' - R_s) = 0 \quad \ldots \ldots \quad (3.20)$$

Eq. 3.20 can be solved by trial and error to find the particular oil bank pressure required to dissolve the trapped gas.

The preceding derivation assumes that the gas solubility and oil reservoir volume factor data obtained for use with declining pressures can also be used to represent events occurring with increasing pressure. Typically, an increase in pressure of 10 to 20 percent is enough to dissolve any gas trapped in the oil bank.

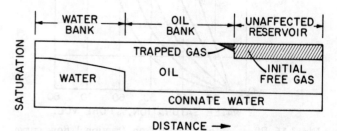

Fig. 3.20 Saturation profile during a waterflood.

Efficiency of Oil Displacement by Water

Effect of Trapped Gas on Waterflood Recovery

Since 1949 a number of experimental studies[3,36-43] have been made of the effect of an initial gas saturation on waterflood recovery. These studies have included measurements of initial flowing gas saturation, subsequent trapped gas saturations, and residual oil saturations following a waterflood with trapped gas in place. To be meaningful, these studies must be carried out under conditions such that no significant compression or solution of gas occurs during either the trapping of

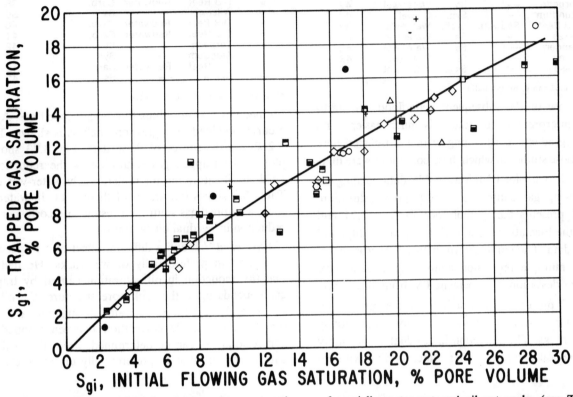

Fig. 3.21 Relation between initial and trapped gas saturation, preferentially water-wet and oil-wet rocks (see Table 3.1 for symbol code).

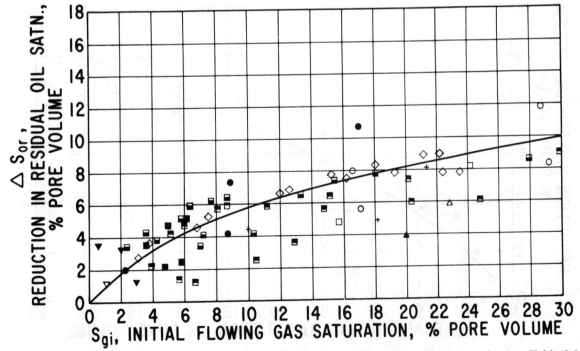

Fig. 3.22 Effect of initial gas saturation on waterflood oil recovery, preferentially water-wet rocks (see Table 3.2 for symbol code).

TABLE 3.1 — RELATION BETWEEN INITIAL AND TRAPPED GAS SATURATION
(See Fig. 3.21)

Symbol	Formation	Field	Rock Type*	Wettability Preference	Source (reference)
●	Nellie Bly	—	Ss	Water-wet	3
◇	Wilcox	—	Ss	Water-wet	39
◨	D-3 Reef	Redwater	Carb.	—	40
◨	D-3 Reef	Redwater	Carb.	—	41
+	Torpedo	—	Ss	Water-wet	42
△	Alundum	—	Syn.	Oil-wet	42
◻	D-3 Reef	Redwater	Carb.	Water-wet	42
⌀	Berea	—	Ss	Water-wet	42
⌑	Bandera	—	Ss	Water-wet	43
◐	Berea	—	Ss	Water-wet	43
◓	Boise	—	Ss	Water-wet	43

*Sandstone, carbonate, or synthetic

TABLE 3.2 — EFFECT OF INITIAL GAS SATURATION ON WATERFLOOD OIL RECOVERY, PREFERENTIALLY WATER-WET ROCK
(See Fig. 3.22)

Symbol	Formation	Field	Rock Type*	Source (reference)
▽	Venango	—	Ss	37
▼	Rushford	—	Ss	37
●	Nellie Bly	—	Ss	3
◨	D-3 Reef	Redwater	Carb.	38
◇	Wilcox	—	Ss	39
◨	D-3 Reef	Redwater	Carb.	40
◨	D-3 Reef	Redwater	Carb.	41
+	Torpedo	—	Ss	42
△	Alundum	—	Syn.	42
◻	D-3 Reef	Redwater	Carb.	42
⌀	Berea	—	Ss	42

*Sandstone, carbonate, or synthetic

the gas phase or the subsequent waterfloods. In addition, the interpretation of results is much easier if no saturation gradients exist during testing. Discussed here are only those studies in which these conditions were met.

Fig. 3.21 is a plot of the available data relating an initial flowing gas saturation to a trapped saturation. Table 3.1 presents the symbol code for Fig. 3.21 and indicates the formation, field, rock type, wettability, and source of data. The curve shown in Fig. 3.21 was determined by multiple regression analysis. The data have an average deviation of 1 percent PV from this curve.

Fig. 3.22 presents a plot of initial gas saturation on the reduction in residual oil saturation for preferentially water-wet rocks. The reduction in residual oil is simply the residual oil saturation after waterflooding with no trapped gas present minus that with trapped gas. Again, a curve obtained by regression analysis is shown. Table 3.2 shows the source of these data. Fig. 3.23 shows a plot of the trapped gas saturation vs the reduction in residual oil saturation for preferentially water-wet rock. Table 3.3 shows the source of these data. In water-wet rock, a trapped gas saturation results in a corresponding reduction in residual oil saturation.

Kyte et al.[42] found a different effect of a trapped gas saturation in preferentially oil-wet rocks. He deduced that the reduction in oil saturation caused by trapped gas depends upon the rock pore structure, the oil viscosity, and the volume of water throughput, and that, therefore, no simple correlation between trapped gas and residual oil can be determined. From the available data it appears that in preferentially oil-wet rocks, the

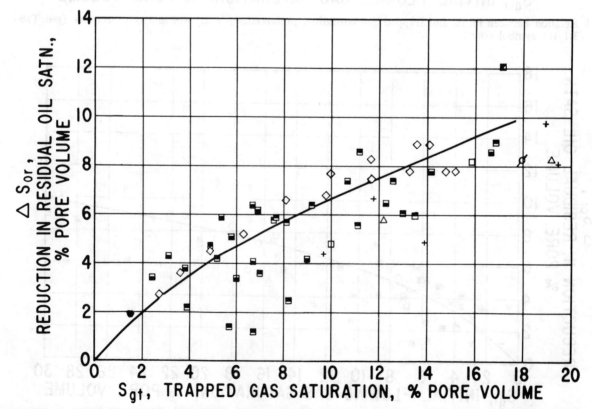

Fig. 3.23 Effect of trapped gas saturation on waterflood recovery, preferentially water-wet rock (see Table 3.3 for symbol code).

EFFICIENCY OF OIL DISPLACEMENT BY WATER

TABLE 3.3 — EFFECT OF TRAPPED GAS SATURATION ON WATERFLOOD OIL RECOVERY, PREFERENTIALLY WATER-WET ROCK
(See Fig. 3.23)

Symbol	Formation	Field	Rock Type*	Source (reference)
●	Nellie Bly	—	Ss	3
◊	Wilcox	—	Ss	39
▫	D-3 Reef	Redwater	Carb.	40
▪	D-3 Reef	Redwater	Carb.	41
+	Torpedo	—	Ss	42
△	Alundum	—	Syn.	42
□	D-3 Reef	Redwater	Carb.	42
▽	Berea	—	Ss	42

*Sandstone, carbonate, or synthetic

oil remaining after the injection of a large number of pore volumes of water does not depend upon whether an initial gas saturation is present. The effect of an initial gas saturation is to reduce the injected water volume required to attain any oil recovery (Fig. 3.24). This results in a reduced producing WOR at any oil recovery when trapped gas is present. The waterflood performance resulting from trapped gas is thus much the same as that obtained with water of increased viscosity.

References

1. Leverett, M. C.: "Capillary Behavior in Porous Solids", *Trans.*, AIME (1941) **142**, 152-169.

2. Buckley, S. E. and Leverett, M. C.: "Mechanism of Fluid Displacements in Sands", *Trans.*, AIME (1942) **146**, 107-116.

3. Holmgren, C. R. and Morse, R. A.: "Effect of Free Gas Saturation on Oil Recovery by Waterflooding", *Trans.*, AIME (1951) **192**, 135-140.

4. Terwilliger, P. L., Wilsey, L. E., Hall, H. N., Bridges, P. M. and Morse, R. A.: "An Experimental and Theoretical Investigation of Gravity Drainage Performance", *Trans.*, AIME (1951) **192**, 285-296.

5. Welge, H. J.: "A Simplified Method for Computing Oil Recovery by Gas or Water Drive", *Trans.*, AIME (1952) **195**, 91-98.

6. Kern, L. R.: "Displacement Mechanism in Multi-Well Systems", *Trans.*, AIME (1952) **195**, 39-46.

7. Martin, J. C.: "Some Mathematical Aspects of Two-Phase Flow With Application to Flooding and Gravity Segregation Problems", *Prod. Monthly* (April, 1958) **23**, No. 4, 22-35.

8. Efros, D. A.: "The Displacement of a Two-Component Mixture When the Viscosity of One of the Fluids Being Displaced Is Low", *Dokladi Akad. Nauk SSR* (July-Aug., 1958) **121**, 59-62.

9. McEwen, C. R.: "A Numerical Solution of the Linear Displacement Equation With Capillary Pressure", *Trans.*, AIME (1959) **216**, 412-415.

10. Sheffield, M.: "Three-Phase Fluid Flow Including Gravitational, Viscous, and Capillary Forces", *Soc. Pet. Eng. J.* (June, 1969) 255-269.

11. Cardwell, W. T., Jr.: "The Meaning of the Triple Value in Non-Capillary Buckley-Leverett Theory", *Trans.*, AIME (1959) **216**, 271-276.

12. Fayers, F. J. and Sheldon, J. W.: "The Effect of Capillary Pressure and Gravity on Two-Phase Fluid Flow in a Porous Medium", *Trans.*, AIME (1959) **216**, 147-155.

13. Hovanessian, S. A. and Fayers, F. J.: "Linear Water Flood with Gravity and Capillary Effects", *Soc. Pet. Eng. J.* (March, 1961) 32-36.

14. Douglas, J., Jr., Blair, P. M. and Wagner, R. J.: "Calculation of Linear Waterflood Behavior Including the Effects of Capillary Pressure", *Trans.*, AIME (1958) **213**, 96-102.

15. Gottfried, B. S., Guilinger, W. H. and Snyder, R. W.: "Numerical Solutions for the Equations for One-Dimensional Multi-Phase Flow in Porous Media", *Soc. Pet. Eng. J.* (March, 1966) 62-72.

16. Levine, J. S.: "Displacement Experiments in a Consolidated Porous System", *Trans.*, AIME (1954) **201**, 57-66.

17. Owens, W. W., Parrish, D. R. and Lamoreaux, W. E.: "An Evaluation of a Gas Drive Method for Determining Relative Permeability Relationships", *Trans.*, AIME (1956) **207**, 275-280.

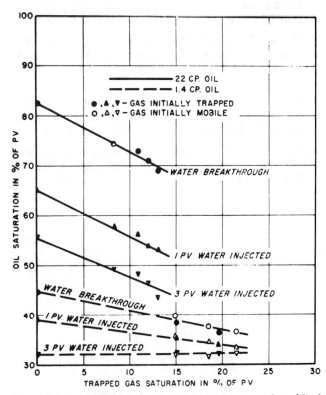

Fig. 3.24 Relationships between trapped gas and residual oil saturations. Dri-filmed alundum core (DFA87).[42]

18. Johnson, E. F., Bossler, D. P. and Naumann, V. O.: "Calculation of Relative Permeability from Displacement Experiments", *Trans.*, AIME (1959) **216**, 370-372.

19. Richardson, J. G.: "The Calculation of Waterflood Recovery from Steady State Relative Permeability Data", *Trans.*, AIME (1957) **210**, 373-375.

20. Rapoport, L. A. and Leas, W. J.: "Properties of Linear Waterfloods", *Trans.*, AIME (1953) **189**, 139-148.

21. Jones-Parra, J. and Calhoun, J. C., Jr.: "Computation of a Linear Flood by the Stabilized Zone Method", *Trans.*, AIME (1953) **189**, 335-338.

22. Bail, P. T.: "The Calculation of Water Flood Performance for the Bradford Third Sand from Relative Permeability and Capillary Pressure Data", *Prod. Monthly* (July, 1956) **21**, No. 7, 20-27.

23. Dietz, D. N.: "A Theoretical Approach to the Problem of Encroaching and Bypassing Edge Water", *Proc.*, Koninkl. Ned. Akad. Wetenschap (1953) B56, 38.

24. Croes, G. A. and Schwarz, N.: "Dimensionally Scaled Experiments and the Theories on the Water-Drive Process", *Trans.*, AIME (1955) **204**, 35-42.

25. Engelberts, W. L. and Klinkenberg, L. J.: "Laboratory Experiments on the Displacement of Oil by Water from Packs of Granular Materials", *Proc.*, Third World Pet. Cong. (1951) **II**, 544.

26. van Meurs, P.: "The Use of Transparent Three-Dimensional Models for Studying the Mechanism of Flow Processes in Oil Reservoirs", *Trans.*, AIME (1957) **210**, 295-301.

27. van Meurs, P. and van der Poel, C.: "A Theoretical Description of Water-Drive Processes Involving Viscous Fingering", *Trans.*, AIME (1958) **213**, 103-112.

28. Chuoke, R. L., van Meurs, P. and van der Poel, C.: "The Instability of Slow Immiscible Viscous Liquid-Liquid Displacements in Permeable Media", *Trans.*, AIME (1959) **216**, 188-194.

29. Scheidegger, A. and Johnson, E. F.: "The Statistical Behavior of Instabilities in Displacement Processes in Porous Media", *Cdn. J. Phys.* (1961) **39**, 326-334.

30. Pottier, J. and Jacquard, P.: "Influence of Capillarity on Unstable Displacement of Immiscible Fluids in Porous Media", *Revue IFP* (April, 1963) **18**, No. 4, 527-540.

31. Rachford, H. H., Jr.: "Instability in Water Flooding Oil from Water-Wet Porous Media Containing Connate Water", *Soc. Pet. Eng. J.* (June, 1964) 133-148.

32. Brown, W. O.: "The Mobility of Connate Water During a Water Flood", *Trans.*, AIME (1957) **210**, 190-195.

33. Kelley, D. L. and Caudle, B. H.: "The Effect of Connate Water on the Efficiency of High Viscosity Waterfloods", *J. Pet. Tech.* (Nov., 1966) 1481-1486.

34. Mungan, N.: "Interfacial Effects in Immiscible Liquid-Liquid Displacement in Porous Media", *Soc. Pet. Eng. J.* (Sept., 1966) 247-253.

35. Holmgren, C. R.: "Some Results of Gas and Water Drives on a Long Core", *Trans.*, AIME (1948) **179**, 103-118.

36. Geffen, T. M., Parrish, D. R. Haynes, G. W. and Morse, R. A.: "Efficiency of Gas Displacement from Porous Media by Liquid Flooding", *Trans.*, AIME (1952) **195**, 29-38.

37. Schiffman, L. and Breston, J. N.: "The Effect of Gas on Recovery When Water Flooding Long Cores", Report No. 54, Pennsylvania State College Mineral Industries Experiment Station (Oct., 1949).

38. "Final Report — Water Flood Study, Redwater Pool", presented to Oil and Gas Conservation Board of Alberta by Canadian Gulf Oil Co., May 22, 1953.

39. Dyes, A. B.: "Production of Water-Driven Reservoirs Below Their Bubble Point", *Trans.*, AIME (1954) **201**, 240-244.

40. "Laboratory Investigation of the Effect of Free Gas in Waterflooding Redwater Cores", report presented to Oil and Gas Conservation Board of Alberta by Imperial Oil Ltd., Oct., 1955.

41. "Laboratory Investigation of the Effect of Free Gas in Waterflooding Redwater Cores", report presented to Oil and Gas Conservation Board of Alberta by Imperial Oil Ltd., March, 1956.

42. Kyte, J. R., Stanclift, R. J., Jr., Stephan, S. C., Jr., and Rapoport, L. A.: "Mechanism of Water Flooding in the Presence of Free Gas", *Trans.*, AIME (1956) **207**, 215-221.

43. Crowell, D. C., Dean, G. W. and Loomis, A. G.: "Efficiency of Gas Displacement from a Water Drive Reservoir", RI 6735, USBM (1966).

Chapter 4

Mobility Ratio Concept

4.1 Development of the Mobility Ratio Concept

In Darcy's law there is a proportionality factor relating the velocity of a fluid to the pressure gradient. This proportionality factor, termed the mobility of the fluid, is the effective permeability of the rock to that fluid, divided by the fluid viscosity. Thus the water mobility is k_w/μ_w and the oil mobility is k_o/μ_o. The value of mobility is dependent upon the fluid saturation.

In 1937 Muskat[1] first discussed the term that has become known as mobility ratio. Later[2,3] it was used to relate the water mobility in the water-contacted portion of a waterflood to the oil mobility in the oil bank. In his second book,[2] Muskat presented the steady-state pressure distributions for a number of injection-production well arrangements—that is, under conditions of a unit mobility ratio.

Aronofsky,[4] in 1952, was the first to stress the importance of the effect of mobility ratio on the flood patterns during water encroachment. He studied, using potentiometric model and numerical techniques, the effect of mobility ratio on the areal coverage of the water-contacted region at the breakthrough of water to the producing wells. This work was closely followed by many others. In the following chapters, we shall see the influence of mobility ratio on areal, vertical, and volumetric sweep efficiency. It is my experience that a flood's mobility ratio is the most important single characteristic of that flood.

Before 1957, the term "mobility ratio" in any paper appearing in the *Transactions* of the AIME was defined at the author's discretion. As a result, in some papers the mobility ratio is defined as the ratio of the oil to the displacing fluid mobility, and in others it is the ratio of displacing fluid to oil mobility. As a result, anyone reading AIME literature published prior to 1957 is cautioned to be aware of the definition of mobility ratio for each paper. Upon adoption of Standard Letter Symbols for Reservoir Engineering in 1957, "mobility ratio" was defined as

$$M = \frac{k_d}{\mu_d} \frac{\mu_o}{k_o} , \quad \quad \quad \quad (4.1)$$

where the subscript d denotes the displacing phase. In waterflooding terminology, this becomes

$$M = \frac{k_w}{\mu_w} \frac{\mu_o}{k_o} = \frac{k_{rw}}{\mu_w} \frac{\mu_o}{k_{ro}} \quad \quad (4.1a)$$

Unfortunately some engineers recognize the combination of oil and water relative permeabilities and viscosities as being similar in form to that in the denominator of the fractional flow equation. I remind the reader of the difference:

1. In the fractional flow equation, the ratio of the relative permeabilities is the ratio at *one* given saturation—that is, at *one* point in the reservoir.
2. In the mobility ratio equation, the water permeability is that in the water-contacted portion of the reservoir, and the oil permeability is that in the oil bank—that is, at *two* different and separated points in the reservoir.

4.2 Definition of Mobility Ratio

Of course, to use the term "mobility ratio", we must put values to the terms that constitute it. In a waterflood in which there is no saturation gradient behind the waterflood front, there is no ambiguity about the value of water relative permeability to be used. However, in a waterflood—or a gas injection project, for that matter—in which there is a saturation gradient behind the flood front, how can we select the appropriate value of water relative permeability?

Some of the early experimental results avoided that question by the use of potentiometric models or flow models with miscible fluids in which there is no saturation gradient behind the injected fluid front. In 1955, Craig *et al.*[5] presented the results of model waterfloods and gas drives in five-spot patterns. We used a variety of oil viscosities to obtain a range of saturation gradients. We found that if the water mobility were defined at the average water saturation behind the flood front at water breakthrough, the data on areal-sweep:mobility ratio would match those obtained by other investigators[6] using miscible fluids. Fig. 4.1 shows a reproduction of Fig. 2 in Ref. 4, with the mobility ratio term replotted in accordance with the current definition. As we can note, defining the mobility ratio using the displacing fluid mobility at floodout (the right extremity of the horizontal bars) or at the flood front (the left extremity) gives an ambiguous correlation. (The arrows at the ends of the bars are to indicate that the floodout or flood front mobility ratios are off the scale.) As a result of this study, the water mobility is defined as that at the average water saturation in the water-contacted portion of the reservoir. This definition has been widely accepted. Although the displacing-fluid mobility should include the mobility of the movable oil behind the flood front, the discussion of Ref. 4 showed that this was insignificant compared with the water mobility. The relative permeability to oil in the oil bank ahead of the waterflood front is 1.0 in the absence of mobile connate water.

Craig et al.[5] also determined that the average water saturation in the five-spot pattern at water breakthrough is equal to that in a linear system. The average water saturation in the water-contacted portion of a linear system is constant up to breakthrough, but can increase after water breakthrough; the same effect can be noted in a pattern flood. It is correspondingly true that the mobility ratio of a waterflood will remain constant before breakthrough; but after breakthrough it can increase corresponding to the increase in water saturation and water permeability in the water-contacted portion of the pattern. Thus we should keep in mind that there are before-breakthrough and after-breakthrough mobility ratios. In general, if no further designation is applied,

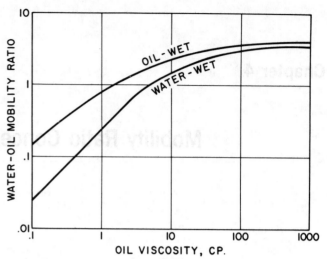

Fig. 4.2 Effect of oil viscosity on water-oil mobility ratio. Water viscosity = 0.5 cp.

the term "mobility ratio" is taken to mean that ratio prior to water breakthrough.

As we shall discuss in later chapters the areal and vertical coverage of a reservoir by any injected fluid is improved at low values of mobility ratio and reduced at high values of mobility ratio. By conventional use, mobility ratios less than unity are termed "favorable", and those greater than unity are "unfavorable". Qualitative phrases such as "very favorable" or "very unfavorable" are used to indicate an increased distance from a unit mobility ratio.

4.3 Ranges of Mobility Ratios During Waterflooding

Based on the preferentially water-wet and oil-wet relative permeability curves shown in Figs. 2.16 and 2.17, the water-oil mobility ratios have been calculated using

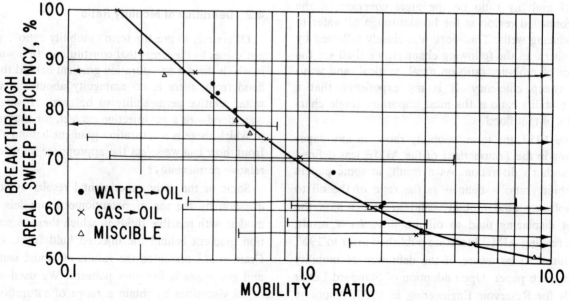

Fig. 4.1 Areal sweep efficiency at breakthrough, five-spot pattern.

a water viscosity of 0.5 cp. These are shown in Fig. 4.2. As the reservoir oil viscosity ranges from 0.1 to 1,000 cp, the waterflood mobility ratio increases from 0.024 to near 3.5 for a water-wet system and from 0.15 to 4.2 for an oil-wet system. The most commonly encountered values of mobility ratios during waterflooding range from 0.02 to 2.0.

The curves shown on Fig. 4.2 are for specific water-oil relative permeability characteristics. Use of these curves is *not* recommended to estimate a value of mobility ratio for other sets of rock flow properties.

References

1. Muskat, M.: *The Flow of Homogeneous Fluids Through Porous Media,* McGraw-Hill Book Co., Inc., New York (1937).

2. Muskat, M.: *Physical Principles of Oil Production,* McGraw-Hill Book Co., Inc., New York (1949).

3. Dykstra, H. and Parsons, R. L.: "The Prediction of Waterflood Performance With Variation in Permeability Profile", *Prod. Monthly* (1950) **15,** 9-12.

4. Aronofsky, J.: "Mobility Ratio, Its Influence on Flood Patterns During Water Encroachment", *Trans.,* AIME (1952) **195,** 15-24.

5. Craig, F. F., Jr., Geffen, T. M. and Morse, R. A.: "Oil Recovery Performance of Pattern Gas or Water Injection Operations from Model Tests", *Trans.,* AIME (1955) **204,** 7-15.

6. Slobod, R. L. and Caudle, B. H.: "X-Ray Shadowgraph Studies of Areal Sweepout Efficiencies", *Trans.,* AIME (1952) **195,** 265-270.

Chapter 5

Areal Sweep Efficiency

5.1 Definition

In waterflooding, water is injected into some wells and produced from other wells. In an areal sense, injection and production take place at points. As a result, pressure distributions and corresponding streamlines are developed between injection and production wells. In symmetrical well patterns, a straight line connecting the injector and producer is the shortest streamline between these two wells, and as a result, the pressure gradient along this line is the highest. So injected water moving areally along this shortest streamline reaches the producing well before the water moving along any other streamline. Therefore at the time of water breakthrough, only a portion of the reservoir area lying between these two wells is contacted by water. This contacted fraction is the pattern areal sweep efficiency at breakthrough, E_{Pbt}.

A wide variety of injection-production well arrangements have received attention in the literature. These are shown in Fig. 5.1. Some of them, such as the two-spot and three-spot, are isolated well arrangements for possible pilot flooding purposes. The rest are largely portions of repeating injection-producing well patterns. Note that the regular four-spot and inverted seven-spot patterns are identical. Table 5.1[1] summarizes characteristics of these dispersed flooding patterns. The patterns termed "inverted" have only one injection well per pattern. This is the difference between "normal" and "inverted" well arrangements.

5.2 Measurement

Techniques

The earliest studies involving injection-production well arrangements were calculations of the steady-state pressure distribution.[2] As an extension of these calculations, the areal sweep efficiency at breakthrough was determined[3-5] for a mobility ratio of unity. Later these mathematical studies were broadened[6,7] to consider the influence of mobility ratios other than unity.

Another early technique for determining areal sweep efficiencies involved the electrolytic model[8,9] in which the steady-state flow was duplicated by the flow of electrical current. Modifications of this are the blotter-type model[10] and the gelatin model.[11]

Potentiometric models[12], also, involved the flow of electrical current. They were used as the first approach in experimentally determining areal sweeps at mobility ratios other than unity.[7] In this combined numerical-experimental technique the equipotential and streamline distributions were determined experimentally at a fixed flood front position, and the incremental advance of the flood front was calculated from these distributions; the steps were then repeated.

In the early 1950's the X-ray shadowgraph technique was developed.[13] An X-ray absorber is dissolved in either the injected or the displaced fluid phase. When exposed to an X-ray beam, the phase containing the absorbent casts a shadow on a piece of film, thus indicating the position of the flood front. The initial application was in the use of miscible fluids,[13] but it was soon applied to immiscible fluid flow[14] as well.

Fluid mappers—the so-called "Hele-Shaw" models[15]—had been used in other fields of science and engineering.[16-18] These models are basically simple, involving two flat pieces of glass separated by only a thin spacer. Because of mixing that occurs during flow, their use is in practice limited to mobility ratios of unity.[19]

TABLE 5.1 — CHARACTERISTICS OF DISPERSED INJECTION PATTERNS[1]

Pattern	Ratio of Producing Wells to Injection Wells	Drilling Pattern Required
Four-spot	2	Equilateral triangle
Skewed four-spot	2	Square
Five-spot	1	Square
Seven-spot	½	Equilateral triangle
Inverted seven-spot (single inj. well)	2	Equilateral triangle
Nine-spot	⅓	Square
Inverted nine-spot (single-inj. well)	3	Square
Direct line drive	1	Rectangle
Staggered line drive	1	Offset lines of wells

AREAL SWEEP EFFICIENCY

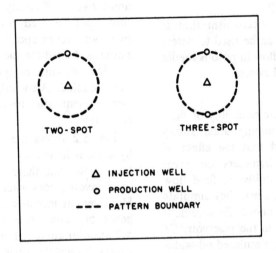
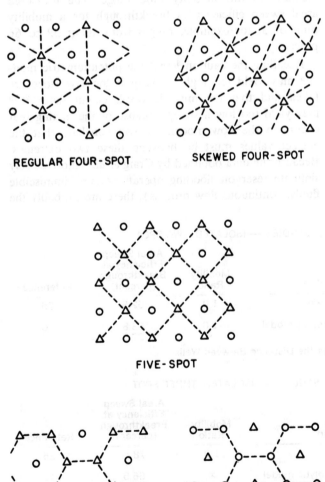
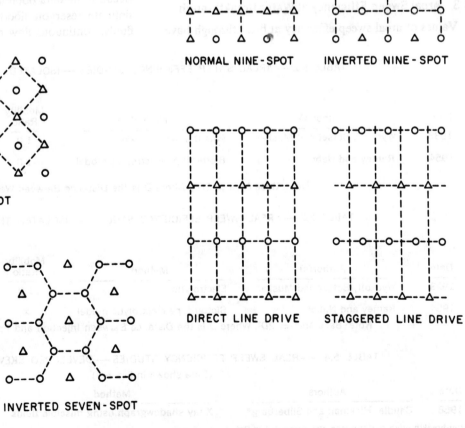

Fig. 5.1 Flooding patterns.

They are a graphic means of depicting the streamline distributions for various well arrangements.[20]

Resistance-type models,[21] simulating a porous medium as a series of interconnected electrical resistances, also were used to obtain areal sweeps.

Thus any physical or chemical mechanism that is governed by the Laplace equation can be used to determine certain characteristics of fluid flow in porous media (obeying Darcy's law) such as areal sweep.

Reliability

The subject of one study[22] was the reliability of fluid flow models as a means of determining areal sweep efficiencies. The authors confirmed that the effect of geometrically unscaled wellbore diameters on areal sweep determinations was small. Intuitively, fluid flow models should be the most reliable, since they simulate in entirety the unsteady-state continuous displacement of one fluid by another that occurs in the reservoir. Of course, stabilized zone effects at the simulated oil-water interface must be scaled.[4] Miscible flood experiments at viscosity ratios above about 2 may be misleading because of fluid mixing at the interface.[23]

The most critical study[23,24] was devoted to the potentiometric model technique. The primary concern was the validity of simulating a continuous displacement process by a small number of discrete steps.[24] A later study concluded that areal sweeps from potentiometric models at high mobility ratios may be optimistic if too few steps are taken.[23]

5.3 Areal Sweep Efficiency at Water Breakthrough

Values of areal sweep efficiency at breakthrough have been reported for a variety of flooding patterns. The most popular pattern for study has been the five-spot pattern.

Tables 5.2 through 5.11 present information on the areal sweep efficiency studies made on the two-spot, three-spot, skewed four-spot, developed and isolated five-spot, seven-spot, nine-spot, and direct-line-drive patterns. Listed for each of these studies are the dates of publication, authors, type of method used, and mobility ratio studied. Also indicated are those references that also present after-breakthrough sweep efficiency performance.

Fig. 5.2 shows the areal sweep efficiencies obtained by various investigators for the developed five-spot pattern. Note that there is satisfactory agreement among most investigators when the mobility ratio is 1.0 or less. However, at mobility ratios above unity a wide divergence between reported values occurs. Shown on Fig. 5.2 are four curves at mobility ratios above unity. These curves follow the various sets of data obtained experimentally in this mobility ratio range. The measured areal sweep efficiency at breakthrough for a mobility ratio of 3, for example, ranges from about 52 to 66 percent.

Recalling the caution[23] that the potentiometric model[28] may yield too high a sweep if too few steps are taken and that the miscible displacement data[13,26,29] may yield too low a sweep because of the mixing that occurs in the flow model, we can conclude that the correct values must be between these two extremes. Because the data obtained by Craig et al.[14] more closely simulate reservoir flooding operations (i.e., immiscible fluids, continuous flow process), these are probably the

TABLE 5.2 — AREAL SWEEP EFFICIENCY STUDIES — ISOLATED TWO-SPOT

Date	Author(s)	Method	Mobility Ratio	Areal Sweep Efficiency at Breakthrough (percent)	Reference
1933	Wyckoff, Botset and Muskat	Potentiometric model	1.0	52.5	25
1954	Ramey and Nabor	Blotter-type electrolytic model	1.0 ∞	53.8 27.7	10

Note: Base Area = $2D^2$, Where D Is the Distance Between Wells

TABLE 5.3 — AREAL SWEEP EFFICIENCY STUDIES — ISOLATED THREE-SPOT

Date	Author(s)	Method	Mobility Ratio	Areal Sweep Efficiency at Breakthrough (percent)	Reference
1933	Wyckoff, Botset and Muskat	Electrolytic	1.0	78.5	25
1954	Ramey and Nabor	Blotter-type electrolytic model	∞	66.5	10

Note: Base Area = $2D^2$, Where D Is the Distance Between Injection and Production Wells

TABLE 5.4 — AREAL SWEEP EFFICIENCY STUDIES — DEVELOPED SKEWED FOUR-SPOT
(Data shown in Fig. 5.3)

Date	Authors	Method	Mobility Ratio	Reference
1968	Caudle, Hickman and Silberberg*	X-ray shadowgraph using miscible fluids	0.1 to 10.0	1

*After-breakthrough performance also presented in Ref. 1.

TABLE 5.5 — AREAL SWEEP EFFICIENCY STUDIES — DEVELOPED FIVE-SPOT
(Data shown in Fig. 5.2)

Date	Author(s)	Method	Mobility Ratio	Reference
1933	Wyckoff, Botset and Muskat	Electrolytic model	1.0	25
1934	Muskat and Wyckoff	Electrolytic model	1.0	8
1951	Fay and Prats	Numerical	4.0	6
1952	Slobod and Caudle	X-ray shadowgraph using miscible fluids	0.1 to 10.0	13
1953	Hurst	Numerical	1.0	5
1954	Dyes, Caudle and Erickson*	X-ray shadowgraph using miscible fluids	0.06 to 10	26
1955	Craig, Geffen and Morse*	X-ray shadowgraph using immiscible fluids	0.16 to 5.0	14
1955	Cheek and Menzie	Fluid mapper	0.04 to 10.0	27
1956	Aronofsky and Ramey	Potentiometric model	0.1 to 10.0	28
1958	Nobles and Janzen	Resistance network	0.1 to 6.0	21
1960	Habermann	Fluid flow model using dyed fluids	0.037 to 130	29
1961	Bradley, Heller and Odeh	Potentiometric model using conductive cloth	0.25 to 4	30

*After-breakthrough performance also presented in these references.

TABLE 5.6 — AREAL SWEEP EFFICIENCY STUDIES — NORMAL AND INVERTED FIVE-SPOT PILOT
(Data shown in Fig. 5.4)

Date	Author(s)	Method	Type	Mobility Ratio	Areal Sweep Efficiency at Breakthrough (percent)	Reference
1958	Paulsell*	Fluid mapper	Inverted	0.319	117.0	19
				1.0	105.0	
				2.01	99.0	
1959	Moss, White and McNiel	Potentiometric	Inverted	∞	92.0	31
1960	Caudle and Loncaric*	X-ray shadowgraph	Normal	0.1 to 10.0	**	32
1962	Neilson and Flock	Rock flow model	Inverted	0.423	110.0	33

*After-breakthrough performance also presented in these references.
**Depends upon the ratio of injection rate to producing rate.

Note: Base Area = D^2, Where D Is the Distance Between Adjacent Producing Wells

TABLE 5.7 — AREAL SWEEP EFFICIENCY STUDIES — DEVELOPED NORMAL SEVEN-SPOT PATTERN
(Data shown in Fig. 5.5)

Date	Author(s)	Method	Mobility Ratio	Areal Sweep Efficiency at Breakthrough (percent)	Reference
1933	Wyckoff, Botset and Muskat	Electrolytic model	1.0	82.0	25
1934	Muskat and Wyckoff	Electrolytic model	1.0	74.0	8
1956	Burton and Crawford*	Gelatin model	0.33	80.5	11
			0.85	77.0	
			2.0	74.5	
1961	Guckert*	X-ray shadowgraph using miscible fluids	0.25	88.1 to 88.2	34
			0.33	88.4 to 88.6	
			0.5	80.3 to 80.5	
			1.0	72.8 to 73.6	
			2.0	68.1 to 69.5	
			3.0	66.0 to 67.3	
			4.0	64.0 to 64.6	

*After-breakthrough performance also presented in these references.

TABLE 5.8 — AREAL SWEEP EFFICIENCY STUDIES — DEVELOPED INVERTED (SINGLE INJECTION WELL) SEVEN-SPOT PATTERN
(Data shown in Fig. 5.6)

Date	Author(s)	Method	Mobility Ratio	Areal Sweep Efficiency at Breakthrough (percent)	Reference
1933	Wyckoff, Botset and Muskat	Electrolytic model	1.0	82.2	25
1956	Burton and Crawford*	Gelatin model	0.5	77.0	11
			1.3	76.0	
			2.5	75.0	
1961	Guckert*	X-ray shadowgraph using miscible fluids	0.25	87.7 to 89.0	34
			0.33	84.0 to 84.7	
			0.50	79.0 to 80.5	
			1.0	72.8 to 73.7	
			2.0	68.8 to 69.0	
			3.0	66.3 to 67.2	
			4.0	63.0 to 63.6	

*After-breakthrough performance also presented in these references.

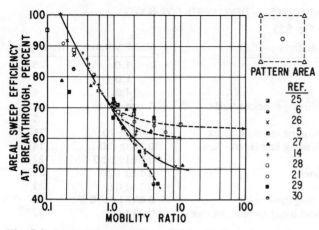

Fig. 5.2 Areal sweep efficiency at breakthrough, developed five-spot pattern.

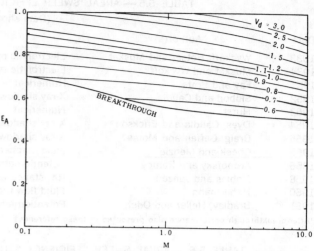

Fig. 5.3 Areal sweep efficiency, skewed four-spot pattern.[1] V_d is the displaceable volume, equal to the cumulative injected fluid as a fraction of the product of the pattern pore volume and displacement efficiency.

most representative of actual waterfloods. (The curve going through the Craig data is the solid line.)

For many of the other patterns studied, too few data are available to yield a plot of any great value.

Fig. 5.3 presents a plot of areal sweep efficiency at breakthrough obtained by Caudle et al.[1] for the skewed four-spot pattern. Also shown on Fig. 5.3 is the areal sweep increase after breakthrough. Fig. 5.4 shows a similar plot for both the single five-spot pattern and the single injection well five-spot pattern—that is, the isolated five-spot pilots, normal and inverted. Figs. 5.5 and 5.6 present the areal sweep efficiency data for the normal and inverted seven-spot patterns, respectively. Two curves are shown on each of these figures, relating two different sets of experimental data. The data of Guckert[34] are recommended for use because of the greater reliability of the X-ray shadowgraph technique at mobility ratios other than unity. The areal sweep efficiency at breakthrough for the inverted nine-spot pattern (one injection well per pattern) described in Table 5.10, depends upon the ratio of the producing rates from the close (side) wells to those from the more distant (corner) producing wells. Areal sweep information[36] for ratios of 0.5, 1.0, and 5.0 are presented in

TABLE 5.9 — AREAL SWEEP EFFICIENCY STUDIES — DEVELOPED NORMAL NINE-SPOT PATTERN

Date	Author	Method	Mobility Ratio	Reference
1939	Krutter	Electrolytic model	1.0	35
1961	Guckert*	X-ray shadowgraph using miscible fluids	1.0 and 2.0	34

*After-breakthrough performance also presented in these references.

TABLE 5.10 — AREAL SWEEP EFFICIENCY STUDIES — INVERTED (SINGLE INJECTION WELL) NINE-SPOT PATTERN
(Data shown in Figs. 5.7 through 5.9)

Date	Author(s)	Method	Mobility Ratio	Reference
1964	Kimbler, Caudle and Cooper*	X-ray shadowgraph using miscible fluids	0.1 to 10.0	36
1964	Watson, Silberberg and Caudle*	Fluid flow model using dyed miscible fluids	0.1 to 10.0	37

*After-breakthrough performance also presented in these references.

TABLE 5.11 — AREAL SWEEP STUDIES — LINE DRIVE PATTERNS

Date	Author(s)	Method	Staggered or Direct Line Drive	d/a	Mobility Ratio	Reference
1933	Wyckoff, Botset and Muskat	Electrolytic model	Direct	1.0	1.0	25
1934	Muskat and Wyckoff	Electrolytic model	Direct Staggered	0.5 to 4.0 0.5 to 4.0	1.0	8
1952	Aronofsky	Numerical and potentiometric model	Direct	1.5	0.1, 1.0, 10	7
1952	Slobod and Caudle	X-ray shadowgraph using miscible fluids	Direct	1.5	0.1 to 10	13
1954	Dyes, Caudle and Erickson*	X-ray shadowgraph using miscible fluids	Direct Staggered	1.0 1.0	0.1 to 17	26
1955	Cheek and Menzie	Fluid mapper	Direct	2.0	0.04 to 11.0	27
1956	Prats	Numerical approach	Staggered	1.0 to 6.0	1.0	38
1962	Burton and Crawford	Gelatin model	Direct	1.0	0.5 to 3.0	11

*After-breakthrough performance also presented in these references.

AREAL SWEEP EFFICIENCY

Figs. 5.7, 5.8, and 5.9, respectively. Fig. 5.10 shows a comparison of the breakthrough areal sweep efficiency at a unit mobility ratio for staggered and direct line drives as a function of d/a, where d is the distance between adjacent rows of wells, and a is the distance between like wells in a row. Two curves are shown on Fig. 5.10 for a staggered line drive pattern, one developed by Muskat[4] and the other by Prats.[38] The Prats results are considered the more reliable. A five-spot pattern is a staggered line drive with a d/a ratio of 0.5. Shown as Figs. 5.11 and 5.12 are the areal sweep efficiencies obtained for both direct and staggered line drives.

Most laboratory pattern sweep studies have involved initially liquid-saturated systems. Areal sweep investigations[14] simulating water injection into a reservoir partially depleted by solution gas drive have indicated that during fillup the water advances radially from the injector. This radial advance continues until either (1) the leading edge of the oil bank contacts an oil bank formed about an adjacent injector or (2) the oil bank encounters a producing well. At that point the water flood front begins to cusp toward the nearest producer. If at this time the flood front should also be radial in the initially liquid-saturated reservoir, the areal sweep at water breakthrough with initial gas *present* would be the same as that with *no* gas. The performance at and after breakthrough would be the same. However, less total oil, by an amount equivalent to the reservoir volume of the initial free gas, will be produced at any injected water volume in the flood following partial depletion. If fillup occurs at a higher areal sweep than that at which radial flow exists, the areal sweep performance with initial gas will be more efficient. For most waterflooding conditions, fillup occurs before the flood fronts cusp, and thus the initial gas saturation does not affect areal sweep or the residual oil saturation.

5.4 Areal Sweep Increase After Water Breakthrough

In his discussion of Aronofsky's paper, Dyes[39] was one of the first to point out that with continued injection after water breakthrough, the areal sweep efficiency of a developed pattern continues to increase until it reaches 100 percent, or complete areal coverage. This has been shown for a four-spot,[1] a developed five-spot,[14,26] a seven-spot,[11,34] a nine-spot,[34,36,37] and for line drives.[26] A number of these correlations are presented with other design charts in Appendix D.

5.5 Injectivities in Various Waterflood Patterns

Of interest, too, in pattern flooding calculations is the water injection rate. Muskat[4] presents steady-state fluid injectivities for a number of different flooding patterns. Deppe[40] presents still others. These are all summarized in Table 5.12. They represent injectivities for steady-state conditions—that is, no initial gas saturation and a mobility ratio of unity.

Several investigators have devoted their attention to the variation in fluid injectivities that might occur in a number of well patterns at mobility ratios other than unity. Three groups of investigators have studied the five-spot pattern, one using potentiometric models,[40] another using a porous flow model and miscible fluids,[41] and the third using resistance networks.[21] All investigators found the same qualitative effect: at favorable mobility ratios ($M < 1$) the fluid injectivity declines as the areal sweep increases, but at unfavorable mobility ratios ($M > 1$) the fluid injectivity increases as complete areal coverage is approached. Of the three groups

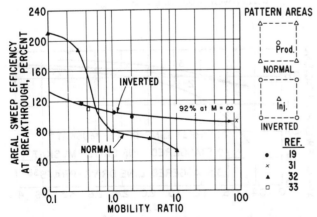

Fig. 5.4 Areal sweep at breakthrough, single injection well (inverted five-spot) and single five-spot (normal five-spot) pilots.

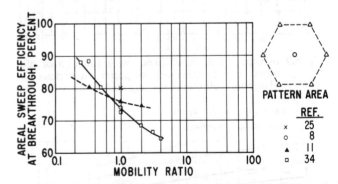

Fig. 5.5 Areal sweep efficiency at breakthrough, developed normal seven-spot pattern.

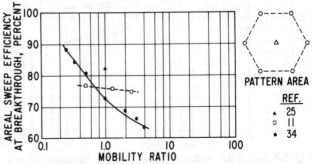

Fig. 5.6 Areal sweep efficiency at breakthrough, developed inverted seven-spot pattern.

of investigators, only Caudle and Witte present data on injectivity variation beyond water breakthrough. Caudle et al.[1] show a similar relationship for a skewed four-spot pattern. In their excellent paper, Prats et al.[42] present a method for predicting injectivity variations in a five-spot pattern having an initial gas saturation.

Let us consider the special case in which there is piston-like displacement—that is, there is no saturation gradient, no flowing oil behind the waterflood front. Then, as discussed in Section 4.2, the water-oil mobility ratio after water breakthrough at the producing wells is equal to that before breakthrough. In this case the injectivity at 100 percent areal coverage is equal to the mobility ratio multiplied by the value calculated from the corresponding equation in Table 5.12.

For reservoir rock-fluid systems in which there is a saturation gradient behind the waterflood front, the mobility ratio will increase following breakthrough. The mobility ratio at floodout, M_{fo}, then is calculated using the relative permeability to water at floodout. Correspondingly the water injectivity at floodout is the steady-state ($M = 1$) injectivity shown in Table 5.12 multiplied by the floodout mobility ratio.

The variation in injectivity with areal sweep during a five-spot waterflood[41] is shown in Fig. 5.13. The same general relationship holds for all patterns. As an approximation, the injection rate at water breakthrough is

$$i_{wbt} = i \times E_{Abt} \times M, \qquad (5.1)$$

where i is the rate calculated from the equation in Table 5.12.

5.6 Areal Sweep Prediction Methods

Some investigators have proposed prediction methods to account specifically for areal sweep and recovery at breakthrough and their increase after breakthrough.

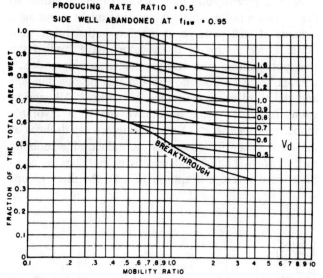

Fig. 5.7 Sweepout pattern efficiency as a function of mobility ratio for the nine-spot pattern at various displaceable volumes (V_d) injected.[36]

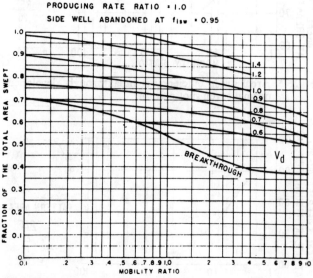

Fig. 5.8 Sweepout pattern efficiency as a function of mobility ratio for the nine-spot pattern at various displaceable volumes (V_d) injected.[36]

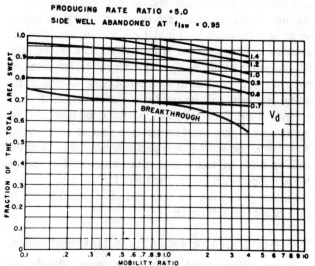

Fig. 5.9 Sweepout pattern efficiency as a function of mobility ratio for the nine-spot pattern at various corner-well producing cuts (f_{icw}).[36]

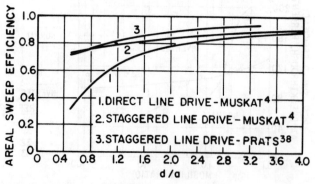

Fig. 5.10 Flooding efficiency of direct line (1) and staggered line drive (2 and 3) well networks as a function of d/a. Mobility ratio = 1.

AREAL SWEEP EFFICIENCY

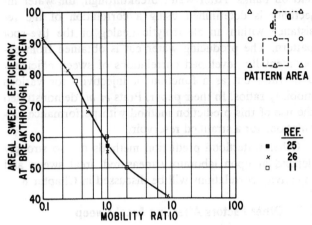

Fig. 5.11 Areal sweep efficiency at breakthrough, developed direct line drive. $d/a = 1.0$.

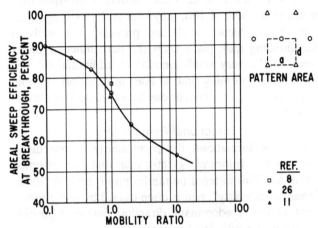

Fig. 5.12 Areal sweep efficiency at breakthrough, developed staggered line drive. $d/a = 1.0$.

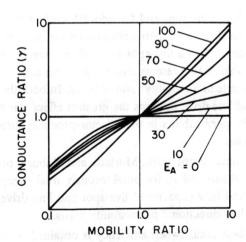

Fig. 5.13 Conductance ratio as a function of mobility ratio and the pattern area mobility ratio and the pattern area swept (E_A), five-spot pattern (after Ref. 41).

Dyes et al. Method

The first of these, Dyes et al.,[26] defined two experimentally determined factors for use in their prediction method:

V_d = displaceable volume, equal to the cumulative injected fluid as a fraction of the product of the pattern pore volume and the displacement efficiency of the flood (this term was earlier[26] designated D_v)

ψ_s = fraction of the total flow coming from the swept portion of the pattern

Plots of V_d and ψ_s as a function of both areal sweep and mobility ratio were presented.[26] For a known displacement efficiency one could select some areal sweep beyond water breakthrough and calculate the following:

1. Total oil recovery as the product of the pattern pore volume, the areal sweep efficiency, and the flood displacement efficiency.

2. Producing WOR from the term ψ_s assuming, for example, that only water is being produced from the swept region. In this event, the water cut would be numerically equivalent to the value of ψ_s.

3. Cumulative injected water volume at that time from V_d by multiplying it by the pattern pore

TABLE 5.12 — INJECTIVITIES FOR REGULAR PATTERNS WITH UNIT MOBILITY RATIO

Direct Line Drive[4] $\left(\dfrac{d}{a} \geq 1\right)$

$$i = \frac{0.001538 \, k \, k_{ro} \, h \, \Delta p}{\mu_o \left(\log \dfrac{a}{r_w} + 0.682 \dfrac{d}{a} - 0.798 \right)}$$

Staggered Line Drive[4] $\left(\dfrac{d}{a} \geq 1\right)$

$$i = \frac{0.001538 \, k \, k_{ro} \, h \, \Delta p}{\mu_o \left(\log \dfrac{a}{r_w} + 0.682 \dfrac{d}{a} - 0.798 \right)}$$

Five-Spot Pattern[4]

$$i = \frac{0.001538 \, k \, k_{ro} \, h \, \Delta p}{\mu_o \left(\log \dfrac{d}{r_w} - 0.2688 \right)}$$

Seven-Spot Pattern[40]

$$i = \frac{0.002051 \, k \, k_{ro} \, h \, \Delta p}{\mu_o \left(\log \dfrac{d}{r_w} - 0.2472 \right)}$$

Inverted Nine-Spot Pattern[40]

$$i = \frac{0.001538 \, k \, k_{ro} \, h \, \Delta p_{i,c}}{\mu_o \left(\dfrac{1+R}{2+R} \right) \left(\log \dfrac{d}{r_w} - 0.1183 \right)}$$

$$i = \frac{0.003076 \, k \, k_{ro} \, h \, \Delta p_{i,s}}{\mu_o \left[\left(\dfrac{3+R}{2+R} \right) \left(\log \dfrac{d}{r_w} - 0.1183 \right) - \dfrac{0.301}{2+R} \right]}$$

R = ratio of producing rates of corner well to side well

$\Delta p_{i,c}$ = pressure differential between injection well and corner producing well

$\Delta p_{i,s}$ = pressure differential between injection well and side producing well

volume, the areal sweep efficiency and finally the flood displacement efficiency.

Necessary laboratory correlations for this approach are available for the five-spot,[26] four-spot,[1] nine-spot,[34] and line drive patterns[26] at a d/a ratio of unity.

These calculated cumulative injected water and produced oil volumes and producing WOR can be converted to a time scale by the use of appropriate fluid injectivity data.[1,41]

Craig et al. Method

Disregarded in the Dyes et al. method was how to handle a flood in which there was a significant amount of producible oil behind the flood front. To solve the problem Craig et al.[14] developed a slightly different prediction technique for a five-spot pattern. They found that the average displacing fluid saturation at breakthrough is the same as that for a linear system. Also the areal sweep efficiency after breakthrough increases directly with the value of log W_i/W_{ibt}, where W_i is the cumulative injected water volume and W_{ibt} is that volume at water breakthrough. They found, too, that the oil and water production performance following water breakthrough can be determined using a modification of Welge's equation.[43] The validity of this calculation method was demonstrated over a range of fluid and rock properties by excellent agreement between calculated and experimental results. Naar et al.[44] confirmed the accuracy of this method for a laboratory five-spot model made up of glass beads.

It should be noted here that at water breakthrough the Craig et al. calculation yields an abrupt increase in water cut from zero to a finite value. There is also a sharp increase late in the flood life to 100 percent water cut. These abrupt changes in predicted water cuts are not realistic but result from approximations used in the correlations. However these sharp changes are less noticeable in multilayer predictions.

Rapoport et al. Method

In 1958, Rapoport et al.[45] introduced what was basically an empirical approach: comparing the waterflood performance of a five-spot pattern with that of a linear system. They presented correlation factors that were dependent upon the oil-water viscosity ratio and, apparently, upon the porous media as well. However, these correlation factors developed by Rapoport et al. from flooding tests using glass beads do not agree with data compiled earlier on consolidated sandstones.[14]

Prats et al. Method

In 1959 Prats et al.[42] presented a combined mathematically-experimentally based procedure for calculating five-spot waterflood performance. Piston-like displacement of oil by water is presumed. Water injectivity into a reservoir that at first is partly depleted by solution gas drive is controlled by the mobility of the water

and oil banks. After water breakthrough, the water injectivity is determined using a correlation of the resistance within an electrolytic analog of the five-spot pattern. The producing water cut is obtained from experimentally developed correlations of sweep efficiency and water cut as a function of injectivity for various mobility ratios. In their paper, Prats et al. demonstrated the use of this prediction method with performance calculations for a stratified reservoir.

Other waterflood prediction methods not so strongly dependent upon laboratory measured areal sweep and injectivity correlations will be discussed in Chapter 8.

5.7 Other Factors Affecting Areal Sweep

Cross-Flooding

Although published laboratory correlations had shown that the areal coverage of a five-spot pattern would increase to 100 percent by continued injection, some still questioned whether cross-flooding would be effective in recovering additional oil. Still and Crawford[46] studied this facet. First flooding a five-spot rock model to a WOR of 100, they shut in the existing injection and production wells and provided new wells so that flooding took place from the perpendicular direction. Only a very small amount of additional oil was produced, confirming that, at least in a uniform permeability system, little benefit can be expected from cross-flooding.

Directional Permeability

In some formations, the permeability in one direction can be significantly greater than that in a direction 90° away. This ordered nonuniformity is termed "directional permeability", and is often caused by water movement in one predominant direction during deposition or during porosity development.

The initial study of the effect of directional permeability on the performance of a five-spot pattern flood was reported by Hutchinson.[47] On the basis of a permeability contrast of 16 to 1, the results shown on Figs. 5.14 and 5.15 were obtained.

In 1960, Landrum and Crawford[48] studied the effect of various directional permeabilities on the flooding performance of both a five-spot and a direct line drive pattern. Table 5.13 presents the results obtained in their study, made at a mobility ratio of unity. In both the five-spot and line drive patterns the greatest effect was noted when the permeability in the injector-producer direction was varied.

In a more recent work, Mortada and Nabor[49] present analytical expressions for breakthrough areal sweep efficiency and flow capacity of five-spot and line drive patterns when directional permeability exists.

The best areal sweep efficiency is obtained when the direction of the maximum permeability is parallel to a line connecting adjacent injection wells.

AREAL SWEEP EFFICIENCY

Permeability Variations

Sandrea and Farouq Ali[50] have turned their attention to the effects of rectilinear impermeable barriers and highly permeable channels present between the injection and production wells in a five-spot. They found that the effect of permeable channels was greater when they existed close to a producing well.

Formation Dip

Prats et al.[51] and Matthews and Fischer[52] studied the effect of formation dip on five-spot waterflood performance at a unit mobility ratio. They found that although the sweep patterns were affected by the dip, the fluid injectivities were not. With the water injection wells moved off pattern in the updip direction, increased sweeps would result at water breakthrough. However these results have little practical significance since few five-spot floods are initiated in dipping reservoirs. When formation dip, flow rate and reservoir permeability are such that gravity effects dominate, it is preferable to inject water along the edge of the pay near the base of the oil column. Gravity will segregate the oil and water and thus allow maximum sweep of the injected water.

Off-Pattern Wells

The effect of off-pattern wells was studied by Prats et al.[53] in 1962. They found that the oil recovery at water breakthrough is always lower with an off-pattern injection well.

When an off-pattern well occurs in a pattern waterflood, for example, the *composite* production performance from all the pattern producers will be close to that for a normal pattern. An early increase in producing WOR will occur at the producer or producers nearest the injector. This will be offset by the late rise in producing WOR at the producer farthest from the injector.

Sweep Beyond Normal Well Pattern

In 1955, Caudle et al.[54] reported on an interesting study of the encroachment of injected water beyond the normal well pattern. This study is applicable to all floods in which a significant volume of formation exists between the last row of wells and the boundary of the reservoir. They found that at least 90 percent of the area lying outside the last row of wells and within one well spacing of these wells would ultimately be swept by the injected water.

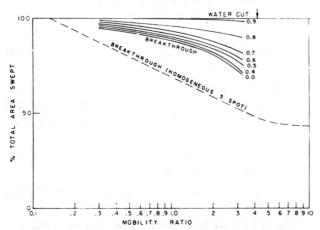

Fig. 5.14 Sweepout pattern efficiency in a five-spot pattern of anisotropic horizontal permeability. The most favorable arrangement has the direction of maximum permeability parallel to lines through injection wells, as illustrated here.[47] Permeability contrast is 16 to 1.

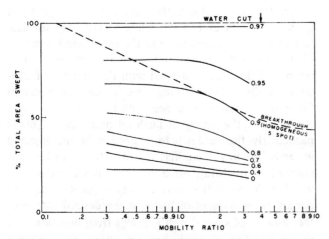

Fig. 5.15 Sweepout pattern efficiency in a five-spot pattern operating under the least favorable arrangement; i.e., with the direction of maximum permeability parallel to a line from an injection well direct to producing well.[47] Permeability contrast is 16 to 1.

TABLE 5.13 — EFFECT OF DIRECTIONAL PERMEABILITY OF AREAL SWEEP EFFICIENCY AT WATER BREAKTHROUGH, MOBILITY RATIO = 1.0
(after Landrum and Crawford[48])

Permeability k_x, Parallel to Line Joining Injector and Producer

	Areal Sweep Efficiency at Breakthrough (percent)	
k_x/k_y	Five-Spot	Direct Line Drive, $d/a = 1.0$
0.01	(1)	100
0.1	12*	95
0.33	(43)	80
1.0	72	56
3.0	43	—
10.0	15*	10
100.0	1	—

Permeability, k_x, 45° from Line Joining Injector and Producer

	Areal Sweep Efficiency at Breakthrough (percent)	
k_x/k_y	Five-Spot	Direct Line Drive, $d/a = 1.0$
0.01	(100)	—
0.1	90	59
0.33	(77)	59
1.0	72	56
3.0	77	—
10.0	90	60
100.0	100	—

* Noted deviation from symmetry
( ) Values estimated from symmetry

These results indicate that much of the recoverable oil in the edge areas of many waterfloods will be produced.

End-to-End Flooding

The patterns previously discussed are all either pilot patterns or developed well arrangements. In 1960 Ferrell et al.[55] discussed what can be called an end-to-end flood, or a form of a peripheral flood pattern. The well arrangement consisted of a single injection well with three offsetting producing wells all in a line with the injection well. Using a single mobility ratio, they studied how continuing production after water breakthrough affects the performance of more distant wells. When producers were shut in at water breakthrough, efficient areal coverage was obtained and less injected water was required to recover the oil. Operators of peripheral waterfloods often use this technique of shutting in producers shortly after water breakthrough and taking the oil production from wells ahead of the flood front.

Horizontal Fractures

Horizontal fractures at either injection wells or producing wells in effect increase the wellbore diameter of these wells. A major effect would be expected on the injectivity. Landrum and Crawford[56] found that fractures as low in radius as 0.04 of the well spacing can double the fluid injectivity of normal-sized wells. It would also be expected that any effect of horizontal fractures on areal sweep would depend upon their radius, with small effects at small radii.[11,57] However, as the radius increases, the areal sweep efficiency can, of course, be drastically reduced.[58]

TABLE 5.14 — EFFECT OF VERTICAL FRACTURES ON FIVE-SPOT PATTERN SWEEP PERFORMANCE; FRACTURES IN LINE WITH INJECTION-PRODUCTION WELL DIRECTION

(After Dyes, Kemp and Caudle[61])

Well Fractured	Fracture Length (fraction of distance between injector and producer)	M	Areal Sweep Efficiency (percent) Breakthrough	90 Percent Water Cut	Throughput at 90 Percent Water Cut (displaceable PV)
Unfractured		0.1	99	99	1.0
		1.1	72	99	1.8
		3.0	56	92	2.2
Injection	¼	0.1	93	98	1.0
	¼	1.1	45	96	1.7
	¼	3.0	39	92	2.2
	½	0.1	88	98	1.1
	½	1.1	37	96	1.8
	½	3.0	28	92	2.7
	¾	0.1	33	97	1.2
	¾	1.1	14	93	2.3
	¾	3.0	10	83	3.8
Production	¼	0.1	78	98	1.1
	¼	1.1	43	95	1.6
	¼	3.0	40	88	1.9
	½	0.1	38	98	1.2
	½	1.1	24	96	1.7
	½	3.0	22	92	2.1
	¾	0.1	18	98	1.8
	¾	1.1	13	94	2.3
	¾	3.0	9	87	3.3

Vertical Fractures

The effect of vertical fractures on sweep performance has been studied for direct line drive patterns[59] as well as for the five-spot pattern.[60-63] Various orientations of vertical fractures were investigated, together with the effects of length, for performance to water breakthrough as well as beyond. Table 5.14 shows the results of one study,[61] indicating that although the effect of vertical fractures on breakthrough sweep is significant regardless of their orientation, a much smaller effect can be noted on areal coverage at 90 percent water cut. Later work[63] confirmed this finding.

5.8 Factors Affecting Selection of Waterflood Pattern

When an engineer plans a waterflood, he has a number of guidelines to follow. The proposed waterflood pattern should fulfill the following:

1. Provide desired oil production capacity.
2. Provide sufficient water injection rate to yield desired oil productivity.
3. Maximize oil recovery with a minimum of water production.
4. Take advantage of known reservoir nonuniformities—i.e., directional permeability, regional permeability differences, formation fractures, dip, etc.
5. Be compatible with the existing well pattern and require a minimum of new wells.
6. Be compatible with flooding operations of other operators on adjacent leases.

The first choice to be made is the flooding pattern—that is, whether the waterflood should be one of a repeating pattern or should attempt to treat the reservoir as a whole, using a peripheral flood, end-to-end flood, down-the-center line of injection wells, or some combination of these. A peripheral flood generally yields the maximum oil recovery with a minimum of produced water. In such a flood, the production of significant quantities of water can be delayed until only the last row of producers remains. On the other hand, because of the unusually small number of injection wells in a peripheral flood as compared with the number of producing wells, it takes a long time for injected water to fill up the reservoir gas space, with the result that there is a delay in the timing of flush oil production. This is particularly the case where a portion of the injected water is lost to the aquifer. Another factor to be considered in deciding on a peripheral waterflood is whether the formation permeability is great enough to permit the movement of water at the desired rate over the distance of several well spacings from injection well to the last line of producers. Of course, the operator of a peripheral waterflood may choose to convert watered-out producers to injection and thus keep the injection wells as close as possible to the waterflood front without bypassing any movable oil. However, moving the location of injection wells frequently requires laying longer surface water lines, so this is discouraged in high-pressure waterfloods. In dipping reservoirs, operators tend to peripherally flood to take maximum advantage of the formation dip in evening out the waterflood front. To summarize, the choice of either a peripheral or a repeating-pattern waterflood is usually made on the basis of the area and dimensions of the reservoir or lease to be flooded, the need for a fast initial oil production response, and the formation dip and permeability.

If the factors weigh in favor of a pattern flood, the engineer then must decide the type of pattern. Where the wells are on square spacing, as is usual, five-spots and nine-spots are the most common flooding patterns. Laboratory experiments have indicated that both of these yield nearly the same oil recovery and WOR performance.[64,65] The choice can be made primarily on the basis of the water-oil mobility ratio, although reservoir heterogeneity is frequently a factor. The mobility ratio is a measure of the injectivity of a well relative to its productivity. At unfavorable mobility ratios ($M > 1$) the water injectivity of an injector exceeds the oil productivity of a producer after fillup, so to balance the desired oil productivity with water injection, a pattern having more producers than injectors is indicated. For favorable mobility ratios, the reverse is true and the recommended pattern should have more injectors than producers. Thus we note that while a mobility ratio less than unity is favorable from sweep aspects, it is unfavorable from an injectivity standpoint.

Table 5.1 indicates that for a normal nine-spot pattern, the ratio of producers to injectors is $\frac{1}{3}$; for a five-spot, the ratio is 1; and for an inverted nine-spot the ratio is 3.

The waterflood operator is generally hesitant to convert a "good" producer to water injection and usually prefers to use a well whose productivity is poorer. Such a decision is unfortunate when the waterflood is operating at a favorable mobility ratio, since the injection rate is thus further impaired. In short, poor producers do not usually make good injection wells, and in fact more frequently make poor injectors.

Where directional permeability or reservoir fractures are known to exist, the prudent engineer will arrange his pattern so that the direction of maximum permeability or orientation of reservoir fractures is in the same direction as the line joining adjacent injectors. As we have noted before, fractures or directional permeability *in the injector-producer direction* results in early water breakthrough and in subsequent large volumes of produced water. Sometimes, the directional movement of water can have disastrous results, as in the case of the North Burbank Unit waterflood.[66,67]

References

1. Caudle, B. H., Hickman, B. M. and Silberberg, I. H.: "Performance of the Skewed Four-Spot Injection Pattern", *J. Pet. Tech.* (Nov., 1968) 1315-1319.

2. Muskat, M.: *Flow of Homogeneous Fluids Through Porous Systems*, J. W. Edwards, Inc., Ann Arbor, Mich. (1946).

3. Muskat, M.: "The Theory of Nine-Spot Flooding Networks", *Prod. Monthly* (March, 1948) **13**, No. 3, 14.

4. Muskat, M.: *Physical Principles of Oil Production*, McGraw-Hill Book Co., Inc., New York (1950).

5. Hurst, W.: "Determination of Performance Curves in Five-Spot Waterflood", *Pet. Eng.* (1953) **25**, B40-46.

6. Fay, C. H. and Prats, M.: "The Application of Numerical Methods to Cycling and Flooding Problems", *Proc.*, Third World Pet. Cong. (1951) **2**, 555-563.

7. Aronofsky, J.: "Mobility Ratio—Its Influence on Flood Patterns During Water Encroachment", *Trans.*, AIME (1952) **195**, 15-24.

8. Muskat, M. and Wyckoff, R. D.: "A Theoretical Analysis of Waterflooding Networks", *Trans.*, AIME (1934) **107**, 62-76.

9. Botset, H. G.: "The Electrolytic Model and Its Application to the Study of Recovery Problems", *Trans.*, AIME (1946) **165**, 15-25.

10. Ramey, H. J., Jr., and Nabor, G. W.: "A Blotter-Type Electrolytic Model Determination of Areal Sweeps in Oil Recovery by In-Situ Combustion", *Trans.*, AIME (1954) **201**, 119-123.

11. Burton, M. B., Jr., and Crawford, P. B.: "Application of the Gelatin Model for Studying Mobility Ratio Effects", *Trans.*, AIME (1956) **207**, 333-337.

12. Lee, B. D.: "Potentiometric Model Studies of Fluid Flow in Petroleum Reservoirs", *Trans.*, AIME (1948) **174**, 41-66.

13. Slobod, R. L. and Caudle, B. H.: "X-Ray Shadowgraph Studies of Areal Sweepout Efficiencies", *Trans.*, AIME (1952) **195**, 265-270.

14. Craig, F. F., Jr., Geffen, T. M. and Morse, R. A.: "Oil Recovery Performance of Pattern Gas or Water Injection Operations from Model Tests", *Trans.*, AIME (1955) **204**, 7-15.

15. Hele-Shaw, H. S.: "Experiments on the Nature of the Surface Resistance in Pipes and on Ships", *Trans.*, Institution of Naval Architects (1897) **XXXIX**, 145.

16. Moore, A. D.: "Fields From Fluid Flow Mappers", *J. Appl. Phys.* (1949) **20**, 790.

17. Moore, A. D.: "The Further Development of Fluid Mappers", *Trans.*, AIEE (1950) **69**, Part II, 1615-1624.

18. Moore, A. D.: "Mapping Technique Applied to Fluid Mapper Patterns", *Trans.*, AIEE (1952) **71**, Part I, 1-4.

19. Paulsell, B. L.: "Areal Sweep Performance of Five-Spot Pilot Floods", MS thesis, The Pennsylvania State U., University Park (Jan., 1958).

20. Henley, D. H.: "Method for Studying Waterflooding Using Analog, Digital, and Rock Models", paper presented at 24th Technical Conference on Petroleum, The Pennsylvania State U., University Park (Oct., 1953).

21. Nobles, M. A. and Janzen, H. B.: "Application of a Resistance Network for Studying Mobility Ratio Effects", *Trans.*, AIME (1958) **213**, 356-358.

22. Slobod, R. L. and Crawford, D. A.: "Evaluation of Reliability of Fluid Flow Models for Areal Sweepout Studies", *Prod. Monthly* (Oct., 1962) **27**, 18-22.

23. Dougherty, E. L. and Sheldon, J. W.: "The Use of Fluid-Fluid Interfaces to Predict the Behavior of Oil Recovery Processes", *Soc. Pet. Eng. J.* (June, 1964) 171-182.

24. Koeller, R. C. and Craig, F. F., Jr.: Discussion of "Mobility Ratio—Its Influence on Injection and Production Histories in Five-Spot Water Flood", *Trans.*, AIME (1956) **207**, 291-292.

25. Wyckoff, R. D., Botset, H. G. and Muskat, M.: "The Mechanics of Porous Flow Applied to Waterflooding Problems", *Trans.*, AIME (1933) **103**, 219-242.

26. Dyes, A. B., Caudle, B. H. and Erickson, R. A.: "Oil Production After Breakthrough — As Influenced by Mobility Ratio", *Trans.*, AIME (1954) **201**, 81-86.

27. Cheek, R. E. and Menzie, D. E.: "Fluid Mapper Model Studies of Mobility Ratio", *Trans.*, AIME (1955) **204**, 278-281.

28. Aronofsky, J. S. and Ramey, H. J., Jr.: "Mobility Ratio —Its Influence on Injection and Production Histories in Five-Spot Water Flood", *Trans.*, AIME (1956) **207**, 205-210.

29. Habermann, B.: "The Efficiency of Miscible Displacement As a Function of Mobility Ratio", *Trans.*, AIME (1960) **219**, 264-272.

30. Bradley, H. B., Heller, J. P. and Odeh, A. S.: "A Potentiometric Study of the Effects of Mobility Ratio on Reservoir Flow Patterns", *Soc. Pet. Eng. J.* (Sept., 1961) 125-129.

31. Moss, J. T., White, P. D. and McNiel, J. S., Jr.: "In-Situ Combustion Process — Results of a Five-Well Field Experiment", *Trans.*, AIME (1959) **216**, 55-64.

32. Caudle, B. H. and Loncaric, I. G.: "Oil Recovery in Five-Spot Pilot Floods", *Trans.*, AIME (1960) **219**, 132-136.

33. Neilson, I. D. R. and Flock, D. L.: "The Effect of a Free Gas Saturation on the Sweep Efficiency of an Isolated Five-Spot", *Bull.*, CIM (1962) **55**, 124-129.

34. Guckert, L. G.: "Areal Sweepout Performance of Seven and Nine-Spot Flood Patterns", MS thesis, The Pennsylvania State U., University Park (Jan., 1961).

35. Krutter, H.: "Nine-Spot Flooding Program", *Oil and Gas J.* (Aug. 17, 1939) **38**, No. 14, 50.

36. Kimbler, O. K., Caudle, B. H. and Cooper, H. E., Jr.: "Areal Sweepout Behavior in a Nine-Spot Injection Pattern", *J. Pet. Tech.* (Feb., 1964) 199-202.

37. Watson, R. E., Silberberg, I. H. and Caudle, B. H.: "Model Studies of Inverted Nine-Spot Injection Pattern", *J. Pet. Tech.* (July, 1964) 801-804.

38. Prats, M.: "The Breakthrough Sweep Efficiency of a Staggered Line Drive", *Trans.*, AIME (1956) **207**, 361-362.

39. Dyes, A. B.: Discussion of "Mobility Ratio, Its Influence on Flood Patterns During Water Encroachment", *Trans.*, AIME (1952) **195**, 22-23.

40. Deppe, J. C.: "Injection Rates—The Effect of Mobility Ratio, Areal Sweep, and Pattern", *Soc. Pet. Eng. J.* (June, 1961) 81-91.

41. Caudle, B. H. and Witte, M. D.: "Production Potential Changes During Sweep-Out in a Five-Spot System", *Trans.*, AIME (1959) **216**, 446-448.

42. Prats, M., Matthews, C. S., Jewett, R. L. and Baker, J. D.: "Prediction of Injection Rate and Production History for Multifluid Five-Spot Floods", *Trans.*, AIME (1959) **216**, 98-105.

43. Welge, H. J.: "A Simplified Method for Computing Oil Recovery by Gas or Water Drive", *Trans.*, AIME (1952) **195**, 91-98.

44. Naar, J., Wygal, R. J. and Henderson, J. H.: "Imbibition Relative Permeability in Unconsolidated Porous Media", *Soc. Pet. Eng. J.* (March, 1962) 13-23.

45. Rapoport, L. A., Carpenter, C. W., Jr., and Leas, W. J.: "Laboratory Studies of Five-Spot Waterflood Performance", *Trans.*, AIME (1958) **213**, 113-120.

46. Still, G. R. and Crawford, P. B.: "Laboratory Evaluation of Oil Recovery by Cross-Flooding", *Prod. Monthly* (Feb., 1963) **28**, 12-19.

47. Hutchinson, C. A., Jr.: "Reservoir Inhomogeneity Assessment and Control", *Pet. Eng.* (Sept., 1959) B19-26.

48. Landrum, B. L. and Crawford, P. B.: "Effect of Directional Permeability on Sweep Efficiency and Production Capacity", *Trans.*, AIME (1960) **219**, 407-411.

49. Mortada, M. and Nabor, G. W.: "An Approximate Method for Determining Areal Sweep Efficiency and Flow Capacity in Formations With Anisotropic Permeability", *Soc. Pet. Eng. J.* (Dec., 1961) 277-286.

50. Sandrea, R. J. and Farouq Ali, S. M.: "The Effects of Isolated Permeability Interferences on the Sweep Efficiency and Conductivity of a Five-Spot Network", *Soc. Pet. Eng. J.* (March, 1967) 20-30.

51. Prats, M., Strickler, W. R. and Matthews, C. S.: "Single-Fluid Five-Spot Floods in Dipping Reservoirs", *Trans.*, AIME (1955) **204**, 160-167.

52. Matthews, C. S. and Fischer, M. J.: "Effect of Dip on Five-Spot Sweep Pattern", *Trans.*, AIME (1956) **207**, 111-117.

53. Prats, M., Hazebroek, P. and Allen, E. E.: "Effect of Off-Pattern Wells on the Behavior of a Five-Spot Flood", *Trans.*, AIME (1962) **225**, 173-178.

54. Caudle, B. H., Erickson, R. A. and Slobod, R. L.: "The Encroachment of Injected Fluids Beyond the Normal Well Pattern", *Trans.*, AIME (1955) **204**, 79-85.

55. Ferrell, H., Irby, T. L., Pruitt, G. T. and Crawford, P. B.: "Model Studies for Injection-Production Well Conversion During a Line Drive Water Flood", *Trans.*, AIME (1960) **219**, 94-98.

56. Landrum, B. L. and Crawford, P. B.: "Estimated Effect of Horizontal Fractures in Thick Reservoirs on Pattern Conductivity", *Trans.*, AIME (1957) **210**, 399-401.

57. Crawford, P. B. and Collins, R. E.: "Analysis of Flooding Horizontally Fractured Thin Reservoirs", *World Oil* (1954) Aug., 139; Sept., 173; Oct., 214; Nov., 212; and Dec., 197.

58. Pinson, J., Simmons, J., Landrum, B. J. and Crawford, P. B.: "Effect of Large Elliptical Fractures on Sweep Efficiencies in Water Flooding or Fluid Injection Programs", *Prod. Monthly* (Nov., 1963) **28**, No. 11, 20-22.

59. Crawford, P. B. and Collins, R. E.: "Estimated Effect of Vertical Fractures on Secondary Recovery", *Trans.*, AIME (1954) **201**, 192-196.

60. Crawford, P. B., Pinson, J. M., Simmons, J. and Landrum, B. L.: "Sweep Efficiencies of Vertically Fractured Five-Spot Patterns", *Pet. Eng.* (March, 1956) **28**, B95-102.

61. Dyes, A. B., Kemp, C. E. and Caudle, B. H.: "Effect of Fractures on Sweep-Out Pattern", *Trans.*, AIME (1958) **213**, 245-249.

62. Simmons, J., Landrum, B. L., Pinson, J. M. and Crawford, P. B.: "Swept Areas After Breakthrough in Vertically Fractured Five-Spot Patterns", *Trans.*, AIME (1959) **216**, 73-77.

63. Hartsock, J. H. and Slobod, R. L.: "The Effect of Mobility Ratio and Vertical Fractures on the Sweep Efficiency of a Five-Spot", *Prod. Monthly* (Sept., 1961) **26**, No. 9, 2-7.

64. Crawford, P. B.: "Laboratory Factors Affecting Water Flood Pattern Performance and Selection", *J. Pet. Tech.* (Dec., 1960) 11-15.

65. Cotman, N. T., Still, G. R. and Crawford, P. B.: "Laboratory Comparison of Oil Recovery in Five-Spot and Nine-Spot Waterflood Patterns", *Prod. Monthly* (Dec., 1962) **27**, No. 12, 10-13.

66. Hunter, Z. Z.: "8½ Million Extra Barrels in 6 Years", *Oil and Gas J.* (Aug. 27, 1956) 92.

67. Hunter, Z. Z.: "Progress Report, North Burbank Unit Water Flood—Jan. 1, 1956", *Drill. and Prod. Prac.*, API (1956) 262.

Chapter 6

Reservoir Heterogeneity

A thorough discourse of the various types of reservoir heterogeneities, their cause, and their measurement would fill a thick volume indeed. So at the outset let us define our objectives. This chapter will be divided into two sections, the first dealing with a general description of all types of reservoir heterogeneities and the second concerned with measures of the degree of vertical permeability stratification. In both these sections the discussion will proceed with the practicing reservoir engineer in mind and wherever possible the emphasis will be upon quantitative measures of reservoir heterogeneity.

6.1 Types of Reservoir Heterogeneities

The geologists tell us that most reservoirs are laid down in a body of water by a long-term process, spanning a variety of depositional environments, in both time and space. As a result of subsequent physical and chemical reorganization, such as compaction, solution, dolomitization, and cementation, the reservoir characteristics are further changed. Thus the heterogeneity of reservoirs is, for the most part, dependent upon the depositional environments and subsequent events, as well as on the nature of particles constituting the sediment. However, we would, in general, expect a reservoir to have some lateral similarity; that is, at an elevation corresponding to a given deposition period, the same basic particle size range should exist over wide areal expanses.[1] The variation in rock properties with elevation would be largely due to differing depositional environments or to segregation of differently sized or constituted sediments into layers, or to both.

In a sandstone reservoir, the development of properties such as porosity and permeability is mostly physical — that is, the properties depend on the nature of the sediment, on the environment of deposition, and generally on subsequent compaction and cementation. In a carbonate reservoir, on the other hand, the development of porosity is more complex. In addition to forming in the same manner as it does in sandstones, carbonate porosity may develop after consolidation or deposition through selective solution, replacement, recrystallization, dolomitization, etc.

In both carbonate and sandstone reservoirs, gross rock movements can result in faulting and, even more important to the reservoir engineer, in the development of both large and small reservoir fractures.

Our discussion of the types of reservoir heterogeneities will be divided into three categories: areal variations, vertical variations, and reservoir-scale fractures. It is obvious that the reservoir may be nonuniform in all intensive properties such as permeability, porosity, pore size distribution, wettability, connate water saturation and crude properties. However, we will primarily discuss the most important of these factors: permeability.

Areal Permeability Variations

Since the early stages of oil production, engineers have recognized, although the fact is sometimes obscured by the effect of different well completion techniques, that most reservoirs vary in permeability in the lateral direction. The first attempt to quantify these areal permeability distributions from observed differences in well production history was that of Kruger[2] in 1961. Using a mathematical model described by McCarty and Barfield,[3] he developed and illustrated a numerical technique. Others[4,5] continued the development of this approach, the latter[5] developing a regression analysis technique for determining a two-dimensional reservoir description from well pressure interference tests. All of these techniques require an electric analyzer or digital computer to handle the time-consuming calculations.

Arnold et al.[6] and Greenkorn et al.[7] devoted their attention to directional permeability effects — that is, permeability anisotropy. Using both pressure data from surrounding wells and core sample permeabilities, techniques were demonstrated for determining the direction and degree of directional permeability. These analyses also require the use of digital computers.

Groult et al.[8] suggested techniques for describing both lateral and vertical inhomogeneities from observations at the formation outcrop and by production logging techniques. Perhaps the simplest approach suggested to date is that described by Johnson et al.[9] and termed "pulse testing". In this procedure a series of producing rate changes or pulses is made at one well with the response being measured at adjacent wells by a differential pressure gauge having a sensitivity of about 0.001 psi. This technique shows promise for providing a measure of the formation flow capacity (kh) and storage capacity (ϕh). In addition, the method can be used qualitatively to measure communication across faults and between zones as well as the direction and magnitude of fracture trends.

A variety of pressure transient techniques has been suggested to provide a measure of (1) the distance to a fault or other impermeable barrier, (2) lateral permeability variations, and (3) the presence, direction and magnitude of natural fracture systems. For a thorough discussion of these, refer to the first Monograph[10] in this series.

Vertical Permeability Stratification

In his 1959 paper, Hutchinson[11] presents an excellent discussion on reservoir nonuniformities. In the section dealing with stratified formations he traces the growth in the concept of layered reservoirs. The attraction of the layered reservoir concept is twofold: it is readily visualized and its reservoir engineering treatment is relatively simple.

In 1963, Elkins and Skov[12] showed that the concept of parallel-layer flow could be used to match the past performance of two gas condensate cycling projects and an enriched gas drive project. Bennion and Griffiths[13] and Testerman[14] have discussed the concept of reservoir stratification and have developed techniques for determining the best description of stratification properties. These will be discussed in detail in a later portion of this chapter.

Several authors[15,16] have suggested that formation outcrops be examined to obtain information on the degree of stratification, lateral extent of shale breaks and continuity of zones of specific permeability. This is no doubt an excellent way for an engineer to actually see the type of formation he is flooding. However, its quantitative usefulness is doubtful. One can never be sure that the depositional environment and subsequent porosity change in the actual reservoir were duplicated in the outcrop portion of the formation.

Reservoir-Scale Fractures and Directional Permeability

Reservoir fractures or closed fracture planes are not uncommon in oil reservoirs. The Spraberry Trend in West Texas is typical of formations so thoroughly penetrated by reservoir-scale fractures that their presence and effect are obvious. Elkins and Skov[17] inferred the orientation of these fractures from pressure transient analysis. Aerial photography also can be helpful here.[18,19] Reservoir engineers recognize that fractures of this type would have an overpowering effect on any attempted waterflood. The engineer should also be cognizant that reservoirs having little indication of fractures during primary depletion may have incipient fractures or "planes of weakness" that manifest themselves when water injection pressure is applied.[20] Their effect can be just as severe as the effect of Spraberry-type fractures. The preponderance of evidence shows that these fractures are not horizontal but generally have a near-vertical orientation, so that they can present highly directional short circuits for the injected water to bypass the oil in much of the matrix rock.

Although directional permeability has been discussed for many years, its effect is generally small in comparison with that of regional variations in permeability, or "permeability trends". The effect of directional permeability can frequently be neglected for practical purposes.

6.2 Quantitative Descriptions of Permeability Stratification

Conformance Factor

One of the earliest measures of reservoir nonuniformities was termed "conformance". Introduced by Patton[21] in 1947, it represents the portion of the reservoir contacted by the injected fluid and, as such, combines areal sweep and vertical sweep effects. The term still finds some use today when engineers use it qualitatively to describe reservoirs as "high conformance" or "low conformance" and thus to indicate the degree of areal and vertical permeability variations. In this sense, conformance implies the fraction of the idealized performance realized.

Positional Approach

In 1947, Miller and Lents[22] presented a means of using core permeabilities for determining layer properties. This approach we have termed the "positional approach". Core data from each well in the Bodcaw Reservoir, Cotton Valley Field Cycling Project were plotted as a function of percent of sand thickness, and the permeabilities were averaged at each percent of sand thickness. This method is equivalent to dividing each well's core analysis into a specified number of vertical segments and determining for each segment in all wells the average permeability and thickness.

Miller and Lents reported that the gas cycling performance of the Bodcaw reservoir agreed closely with that calculated using this layering technique. Elkins and Skov[12] also report success using this approach in match-

TABLE 6.1 — CORE ANALYSIS FOR HYPOTHETICAL RESERVOIR
Cores from 10 Wells, A Through J; Each Permeability Value (md) Represents 1 ft of Pay

Depth (ft)	A	B	C	D	E	F	G	H	I	J
6,791	2.9	7.4	30.4	3.8	8.6	14.5	39.9	2.3	12.0	29.0
6,792	11.3	1.7	17.6	24.6	5.5	5.3	4.8	3.0	0.6	99.0
6,793	2.1	21.2	4.4	2.4	5.0	1.0	3.9	8.4	8.9	7.6
6,794	167.0	1.2	2.6	22.0	11.7	6.7	74.0	25.5	1.5	5.9
6,795	3.6	920.0	37.0	10.4	16.5	11.0	120.0	4.1	3.5	33.5
6,796	19.5	26.6	7.8	32.0	10.7	10.0	19.0	12.4	3.3	6.5
6,797	6.9	3.2	13.1	41.8	9.4	12.9	55.2	2.0	5.2	2.7
6,798	50.4	35.2	0.8	18.4	20.1	27.8	22.7	47.4	4.3	66.0
6,799	16.0	71.5	1.8	14.0	84.0	15.0	6.0	6.3	44.5	5.7
6,800	23.5	13.5	1.5	17.0	9.8	8.1	15.4	4.6	9.1	60.0

ing the performance of a number of fluid injection projects.

To illustrate the use of this technique as well as other measures of permeability stratification, consider the core analysis permeabilities for a hypothetical reservoir shown in Table 6.1. The table shows 10 wells, A through J, with 10 values of permeability per well, each value representing 1 ft of pay.

Arranging these values in order from maximum to minimum, we can obtain the plot shown in Fig. 6.1. This plot relates the portion of the cumulative flow capacity to the portion of the total formation thickness. It is used widely to indicate the contrast in permeabilities, the greater contrast indicated by the increased divergence from a 45° line.

Shown on Table 6.2 are the average layer permeabilities determined for the hypothetical reservoir by the positional approach. These average permeabilities are obtained by taking the geometric average of the permeabilities in each row (i.e., at each depth). This approach has the advantage that it considers both the permeability and the location of a rock sample in determining layer properties.

In Table 6.2, the layers have equal thickness. There are advantages, however, to selecting layer properties so that each layer has the same permeability-thickness product. This will be discussed in Section 7.7.

Coefficient of Permeability Variation

Law[23] showed that rock permeabilities usually have a log normal distribution. This means that plotting the number of samples in any permeability range against the values of log permeability will yield the familiar bell-shaped curve. Fig. 6.2 shows this distribution for the permeability values listed in Table 6.1.

In the first use of core analysis data to measure the effect of permeability stratification on waterflood predictions, Dykstra and Parsons[24] made use of the commonly found log normal permeability distribution of reservoir rock. Their term "coefficient of permeability variation" is frequently shortened to simply "permeability variation". Statistically, the coefficient of variation, V, is defined as

$$V = \frac{\sigma}{\overline{X}}, \qquad (6.1)$$

where

σ = standard deviation
$\overline{X}$ = mean value of X

In a normal distribution the value of σ is such that 15.9 percent of the samples have values of X less than $(\overline{X} - \sigma)$ and 84.1 percent of the samples have values of X less than $(\overline{X} + \sigma)$.

Dykstra and Parsons proposed that permeability values taken from core analyses be arranged in descending order. The percent of the total number of permeability

TABLE 6.2 — LAYER PROPERTIES BY PERMEABILITY ORDERING AND POSITIONAL APPROACH

	Average Layer Permeability (md)	
Layer	Permeability Ordering	Positional Approach
1	84.0	10.0
2	37.0	6.8
3	23.5	4.7
4	16.5	10.4
5	12.0	20.5
6	8.9	12.1
7	6.5	8.6
8	4.6	18.4
9	3.0	14.3
10	1.5	10.9

Arithmetic average permeability   28.2 md
Mean permeability   10.0 md
Ratio of maximum to minimum layer permeability:
  Permeability ordering   84.0/1.5 = 56.0
  Positional approach   20.5/4.7 = 4.37

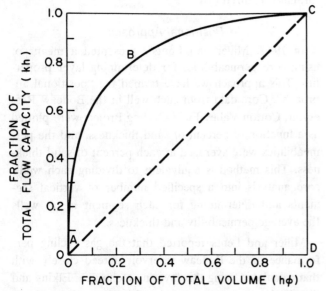

Fig. 6.1 Flow capacity distribution, hypothetical reservoir.

RESERVOIR HETEROGENEITY

values exceeding each tabulated entry is computed. These values are then plotted on log probability paper (Fig. 6.3). The best straight line is drawn through the points, with the central points weighted more heavily than the more distant points. The permeability variation is then

$$V = \frac{\bar{k} - k_\sigma}{\bar{k}}, \qquad (6.2)$$

where

$\bar{k}$ = mean permeability = permeability value with 50 percent probability

k_σ = permeability at 84.1 percent of the cumulative sample.

The possible values of permeability variation range from zero to one, with a completely uniform system having a value of zero.

For those mathematically inclined readers, we should point out that in the true statistical sense, Eq. 6.2 is incorrect. It should read

$$V = \frac{\log \bar{k} - \log k_\sigma}{\log \bar{k}} \qquad (6.3)$$

However, reservoir engineering usage has been with Eq. 6.2.

Dykstra and Parsons went on to correlate their value of permeability variation with expected waterflood performance. This performance prediction technique will be discussed in Chapter 8.

Permeability Ordering

In an early use of core analysis data, Stiles[25] arranged the permeabilities in order from maximum to minimum and then used this distribution in some waterflooding calculations. To differentiate between the method of treating core permeabilities and the performance calculation method, we shall call the former "permeability ordering" and the latter "the Stiles method".

Table 6.2 shows the results of taking the permeability values of Table 6.1, arranging them in order from maximum to minimum, then dividing them in order into 10 equal-sized groups. These groups then represent the average permeabilities within each of the 10 layers of the reservoir, as determined from the permeability ordering method. These averages are those permeabilities, taken from Fig. 6.3, at 10-percent increments, beginning at 5 percent of the total sample. A less severe contrast in layer permeabilities is noted using the positional approach than using the permeability ordering method.

Lorenz Coefficient

In 1950, Schmalz and Rahme[26] proposed a single term for characterizing the permeability distribution within a pay section. Using Fig. 6.1, they defined the Lorenz coefficient of heterogeneity as

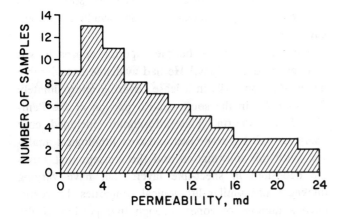

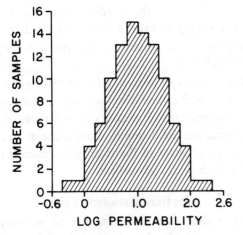

Fig. 6.2 Permeability distribution, hypothetical reservoir.

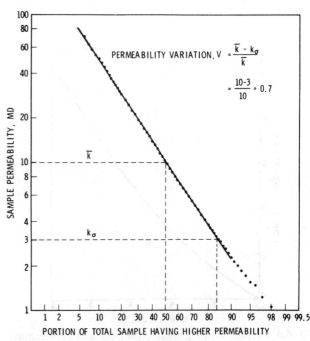

Fig. 6.3 Log normal permeability distribution.

$$\text{Lorenz coefficient} = \frac{\text{area } \overline{ABCA}}{\text{area } \overline{ADCA}}$$

The value of the Lorenz coefficient ranges from zero to 1, a uniform permeability reservoir having a Lorenz coefficient of zero.

The Lorenz coefficient is not a unique measure of reservoir nonuniformity. Several different permeability distributions can yield the same value of Lorenz coefficient.

Fig. 6.4 shows the relation of the permeability variation and Lorenz coefficient for log normal permeability distributions.[27]

Averaging Permeabilities

Warren and Price[27] showed experimentally that the most probable behavior of a heterogeneous system approaches that of a uniform system having a permeability equal to the geometric mean. The geometric mean is:

$$\overline{k} = \sqrt[n]{k_1 \times k_2 \times k_3 \times k_4 \times \ldots \times k_n} \quad . \quad (6.4)$$

It can also be shown analytically that the mean of a log normal distribution is the geometric mean. The geometric mean is the recommended single value of permeability with which to characterize a formation. The value of 10.0 md shown on Table 6.2 is the geometric mean. Calculating the geometric mean (Eq. 6.4) for the permeability values in each layer of Table 6.1, we arrive at the values shown in Table 6.2 for the positional approach.

Permeabilities in series are averaged as follows:

$$\frac{n}{\overline{k}} = \frac{1}{k_1} + \frac{1}{k_2} + \frac{1}{k_3} + \frac{1}{k_4} + \ldots + \frac{1}{k_n}, \quad (6.5)$$

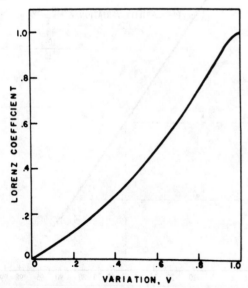

Fig. 6.4 Correlation of Lorenz coefficient and permeability variation.[27]

which assumes that each permeability value represents a unit length.

Permeabilites in parallel are averaged as follows:

$$\overline{k} = \frac{k_1 + k_2 + k_3 + k_4 + \ldots + k_n}{n} \quad . \quad (6.6)$$

Statistical Reservoir Zonation Technique

In 1962, Testerman[14] presented the best available statistical technique for determining layer properties. The technique uses a set of permeability data from a single well showing permeability at various elevations. This set is divided into zones so that the variation in permeability within any one zone is minimized and the contrast between zones is maximized. Statistical criteria are used to determine whether the data will support partitioning into additional zones. These zones are then traced from well to well to obtain a reservoir layer description. This technique has the advantage of providing an unbiased specification of the number and location of reservoir layers, but it does require access to a computer.

Geological Zonation

The previously discussed techniques involve no property of the rock other than permeability. The best method for characterizing the permeability stratification of a reservoir should include also any available geological information.

What I consider to be the superior approach was demonstrated by Alpay.[1] He had available considerable information on wells in a lease covering approximately 7½ sq miles in the south-central portion of the Pembina field, Alberta. This information included core analyses, well log responses and core lithological analyses. With this information and the concept that sands are deposited in sheet-like layers that have varying thickness and similar lithological properties, he correlated a number of zones through that portion of the Cardium sandstone reservoir. A typical result shown in Fig. 6.5 depicts the lateral continuity and varying thickness of a subzone in the Cardium reservoir. Fig. 6.6 shows a stratigraphic breakdown of the Cardium reservoir as given by the gamma ray log, core lithology, and core permeabilities.

Such a study is time-consuming, requiring much detail in core-logging and core-lithology information. It appears that our greatest need today in the area of reservoir performance prediction is a quick, cheap means for obtaining an estimate of the interwell permeability distribution, both areal and vertical.

Crossflow Between Layers

Many methods for predicting the oil recovery performance of waterfloods assume that the layers in the reservoir are each continuous from well to well, uniform in properties, and insulated from each other except at the wellbores. We generally visualize such a reservoir as a layer cake, with icing between each layer serving as the insulating material.

From what we know, very few reservoirs satisfy the concept of shale streaks or impermeable beds acting as the material insulating each layer from the other. Of course, we have known of reservoirs composed of a series of thin sand stringers, each of which could be correlated from well to well. Elkins[12,28] points out that even in clean sands the macroscopic vertical permeability can be several orders of magnitude (100-fold or more) lower than the horizontal permeability. As a result the reservoir can in effect perform as one in which little or no crossflow, or flow between the layers, occurs.

Obviously actual reservoirs will range all the way from those having no crossflow to those having complete vertical transparency to flow. Some reservoir engineers follow the philosophy of considering each reservoir to be composed of insulated layers. As we shall see in the next chapter, this generally provides a conservative estimate of waterflood performance.

References

1. Alpay, O.: "A Study of Physical and Textural Heterogeneity in Sedimentary Rocks", PhD dissertation, Purdue U., Lafayette, Ind. (June, 1963).

2. Kruger, W. D.: "Determining Areal Permeability Distribution by Calculations", *J. Pet. Tech.* (July, 1961) 691-696.

3. McCarty, D. G. and Barfield, E. C.: "The Use of High-Speed Computers for Predicting Flood-Out Patterns", *Trans.*, AIME (1958) **213**, 139-145.

4. Jacquard, P. and Jain, C.: "Permeability Distribution From Field Pressure Data", *Soc. Pet. Eng. J.* (Dec., 1965) 281-294.

5. Jahns, J. O.: "A Rapid Method for Obtaining a Two-Dimensional Reservoir Description from Well Pressure Response Data", *Soc. Pet. Eng. J.* (Dec., 1966) 315-327.

6. Arnold, M. D., Gonzalez, H. J. and Crawford, P. B.: "Estimation of Reservoir Anisotropy from Production Data", *J. Pet. Tech.* (Aug., 1962) 909-912.

7. Greenkorn, R. A., Johnson, C. R. and Stallenberger, L. K.: "Directional Permeability of Heterogeneous Anisotropic Porous Media," *Soc. Pet. Eng. J.* (June, 1964) 124-132.

8. Groult, J., Reiss, L. H. and Montadert, L.: "Reservoir Inhomogeneities Deduced from Outcrop Observations and Production Logging", *J. Pet. Tech.* (July, 1966) 883-891.

9. Johnson, C. R., Greenkorn, R. A. and Woods, E. G.: "Pulse Testing: A New Method for Describing Reservoir Flow Properties Between Wells", *J. Pet. Tech.* (Dec., 1966) 1599-1604.

10. Matthews, C. S. and Russell, D. G.: *Pressure Buildup and Flow Tests in Wells,* Monograph Series, Society of Petroleum Engineers, Dallas, Tex. (1967) **1.**

11. Hutchinson, C. A., Jr.: "Reservoir Inhomogeneity Assessment and Control", *Pet. Eng.* (1959) **31**, B19-26.

12. Elkins, L. F. and Skov, A. M.: "Some Field Observations of Heterogeneity of Reservoir Rocks and Its Effects on Oil Displacement Efficiency", paper SPE 282 presented at SPE Production Research Symposium, Tulsa, Okla., April 12-13, 1962.

13. Bennion, D. W. and Griffiths, J. C.: "A Stochastic Model for Predicting Variations in Reservoir Rock Properties", *Soc. Pet. Eng. J.* (March, 1966) 9-16.

14. Testerman, J. D.: "A Statistical Reservoir Zonation Technique", *J. Pet. Tech.* (Aug., 1962) 889-893.

15. Hutchinson, C. A., Jr., Polasek, T. L., Jr., and Dodge, C. F.: "Identification, Classification and Prediction of Reservoir Nonuniformities Affecting Production Operations", *J. Pet. Tech.* (March, 1961) 223-230.

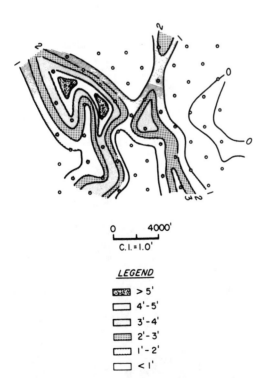

Fig. 6.5 Lateral continuity of Subzone-b, Zone-I portion of Pembina field, Alberta (after Ref. 1).

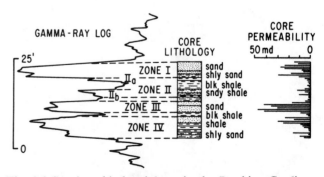

Fig. 6.6 Stratigraphic breakdown in the Pembina Cardium reservoir as given by the gamma ray log, core lithology and core permeability (after Ref. 1).

16. Zieto, G. A.: "Interbedding of Shale Breaks and Reservoir Heterogeneities", *J. Pet. Tech.* (Oct., 1965) 1223-1228.

17. Elkins, L. F. and Skov, A. M.: "Determination of Fracture Orientation from Pressure Interference", *Trans.*, AIME (1960) **219**, 301-304.

18. Donahue, D. A. T.: "Can Aerial Photography Improve Oil Recovery?" *Oil and Gas J.* (July 10, 1967) 201-204.

19. Overbey, W. K., Jr., and Rough, R. L.: "Surface Joint Patterns Predict Well Bore Fracture Orientation", *Oil and Gas J.* (Feb. 26, 1968) 84-86.

20. Clark, K. K.: "Pressure Transient Testing of Fractured Water Injection Wells", *J. Pet. Tech.* (June, 1968) 639-643.

21. Patton, E. C., Jr.: "Evaluation of Pressure Maintenance by Internal Gas Injection in Volumetrically Controlled Reservoirs", *Trans.*, AIME (1947) **170**, 112-155.

22. Miller, M. G. and Lents, M. R.: "Performance of Bodcaw Reservoir, Cotton Valley Field Cycling Project: New Methods of Predicting Gas-Condensate Reservoir Performance Under Cycling Operations Compared to Field Data", *Drill. and Prod. Prac.*, API (1947) 128-149.

23. Law, J.: "Statistical Approach to the Interstitial Heterogeneity of Sand Reservoirs", *Trans.*, AIME (1944) **155**, 202-222.

24. Dykstra, H. and Parsons, R. L.: "The Prediction of Oil Recovery by Water Flood", *Secondary Recovery of Oil in the United States*, 2nd ed., API (1950) 160-174.

25. Stiles, W. E.: "Use of Permeability Distribution in Waterflood Calculations", *Trans.*, AIME (1949) **186**, 9-13.

26. Schmalz, J. P. and Rahme, H. D.: "The Variation of Waterflood Performance With Variation in Permeability Profile", *Prod. Monthly* (1950) **15**, No. 9, 9-12.

27. Warren, J. E. and Price, H. S.: "Flow in Heterogeneous Porous Media", *Soc. Pet. Eng. J.* (Sept., 1961) 153-169.

28. Elkins, L. F.: "Fosterton Field—An Unusual Problem of Bottom Water Coning and Volumetric Water Invasion Efficiency", *Trans.*, AIME (1959) **216**, 130-139.

Chapter 7

Vertical and Volumetric Sweep Efficiencies

7.1 Definition

As a consequence of the nonuniform permeabilities in the vertical dimension, any injected fluid will move as an irregular front. In the more permeable portions of the reservoir, the injected water will travel rapidly, and in the less permeable portions it will move more slowly. A measure of the uniformity of water invasion is the vertical sweep efficiency (designated E_I), also called the invasion efficiency. It is defined as the cross-sectional area contacted by the injected fluid divided by the cross-sectional area enclosed in all layers behind the injected fluid front. The vertical sweep efficiency is a measure of the two-dimensional (vertical cross-section) effect of reservoir nonuniformities.

A closely related term is the volumetric sweep efficiency, E_V, which is a measure of the three-dimensional effect of reservoir heterogeneities. It is equivalent to the product of the pattern areal sweep and vertical sweep; thus

$$E_V = E_P \times E_I \quad \ldots \quad (7.1)$$

The volumetric sweep efficiency can be defined as the pore volume contacted by the injected fluid divided by the total pore volume of a pattern or portion of the reservoir of interest.

In the following sections, we shall discuss the effect of factors such as mobility ratio, gravity and capillary forces, crossflow, and rate on the volumetric sweep efficiency of a waterflood. In these sections we shall avoid discussions of reservoirs with (1) reservoir-scale fractures, (2) areally widespread gas cap, and (3) areally widespread underlying water. Each of these factors represents "short-circuits" for injected water. Waterfloods with any of these conditions present are at a disadvantage at the outset, even though they are sometimes flooded in the hope that additonal profit and oil can be obtained. Though no formal or exact method is available for treating these short-circuits, a thorough understanding of the flow principles in the more tractable geometries treated here can be a real aid to the reservoir engineer's understanding of the problem reservoirs.

7.2 Influence of Mobility Ratio

As previously discussed in Chapter 5, the water-oil mobility ratio is a measure of the water injectivity of a well relative to its oil productivity. That is, after the gas space is filled with liquid, the injectivity variation of a well will depend upon the mobility ratio. This is shown in Fig. 7.1 for a radial system. After fillup, the injectivity will remain constant if the mobility ratio is unity, will increase if $M > 1$, and will decrease if $M < 1$.

During the early period of injection into a reservoir depleted by solution gas drive, the water injectivity declines rapidly from an initial high value. Muskat[1] presented an equation for the injectivity variation during the period of fillup. Expressed in standard SPE symbols, this equation is

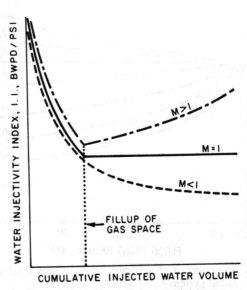

Fig. 7.1 Water injectivity variation, radial system.

$$i_w = \frac{2 \times 10^{-3} \pi hk\, k_{rw}\, (p_{iwf} - p_{ob})}{\mu_w (\ln r_{wo}/r_w + M \ln r_{ob}/r_{wo})}, \quad (7.2)$$

where

p_{ob} = pressure at the leading edge of the oil bank; that is, the initial pressure

r_{wo} = radius of the leading edge of the water bank; that is, at the water-oil interface

r_{ob} = radius of the leading edge of the oil bank.

For a radial system at fillup, then, Eq. 7.2 can be written

$$i_w = \frac{2 \times 10^{-3} \pi hk\, k_{rw}\, (p_{iwf} - p_e)}{\mu_w (\ln r_{wo}/r_w + M \ln r_e/r_{wo})}, \quad (7.3)$$

where subscript e denotes the external boundary of the radial system.

In engineering terms, Eq. 7.3 becomes

$$i_w = \frac{7.07 \times 10^{-3}\, h\, k\, k_{rw}\, (p_{iwf} - p_e)}{\mu_w (\ln r_{wo}/r_w + M \ln r_e/r_{wo})}, \quad (7.3a)$$

where the dimensions are BWPD, ft, md, psi, cp, and ft, respectively.

Fig. 7.2 shows the effect of mobility ratio on the relative injectivity of an initially liquid-filled radial system.[2] The relative injectivity is here defined as the ratio of the injectivity index at any time to that at the start of injection in a liquid-filled reservoir. At a mobility ratio of unity, the relative injectivity is constant. At favorable mobility ratios, $M<1$, the injectivity decreases with an increasing flood-front radius; at unfavorable mobility ratios, $M>1$, the injectivity gradually increases. At the time the waterflood front has reached the external boundary of the radial system, the relative injectivity equals the mobility ratio.

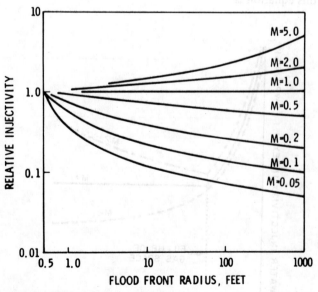

Fig. 7.2 Variation in fluid injectivity in liquid-saturated radial system.[2]
Well radius = 0.5 ft; outer radius = 1,000 ft.

RESERVOIR ENGINEERING ASPECTS OF WATERFLOODING

Caudle and Witte[3] presented a relationship for the injectivity variation with areal coverage in a developed five-spot pattern (Fig. 5.10). Fig. 7.3 shows this relationship plotted in a different form for a 40-acre 5-spot pattern and 1-ft-diameter wellbores. Note here that at 100 percent areal sweep, the relative injectivity is equal to the mobility ratio.

To illustrate how the mobility ratio affects the vertical sweep of both a radial and a five-spot system, let us consider water injection into a liquid-filled formation composed of two layers, identical in thickness and porosity, but having a tenfold contrast in permeability. Fig. 7.4 shows the ratio of the injectivity in the more

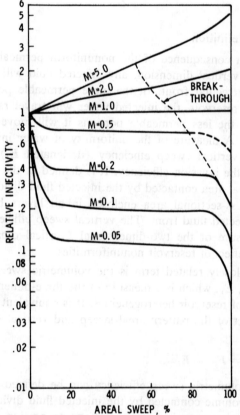

Fig. 7.3 Variation in fluid injectivity in liquid-saturated five-spot.[2] 40-acre five-spot, 1-ft-diameter wellbore.

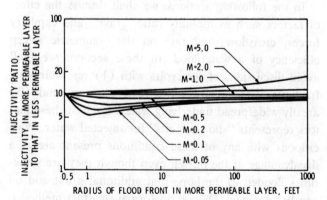

Fig. 7.4 Fluid injectivity contrast in two-layered, liquid-filled radial system.[2]
$r_e = 1,000$ ft; $r_w = 0.5$ ft; $k_1 = 100$ md, $k_2 = 10$ md.

permeable layer to that in the less permeable layer for the radial system as it varies with the radius of the flood front in the more permeable layer. Initially, of course, the injectivity ratio is 10, reflecting the different layer permeabilities. At a unit mobility ratio the injectivity ratio is constant. At unfavorable mobility ratios, $M>1$, the injectivity ratio increases constantly. However, at favorable mobility ratios, $M<1$, the injectivity ratio initially declines, reaches a minimum, then increases gradually. Fig. 7.4 shows that the injectivity ratio at any flood front radius differs only by a factor of about two with mobility ratios from 0.05 to 5.0.

Fig. 7.5 shows the calculated injectivity ratio for a two-layered five-spot pattern, the layer permeabilities differing by a factor of 10. In Fig. 7.5, the wellbore diameter is 1 ft and the pattern area is 40 acres. At unfavorable mobility ratios, $M>1$, the injectivity contrast continually increases with areal sweep to breakthrough. At favorable mobility ratios, $M<1$, the injectivity ratio declines sharply at low areal sweeps, then increases to a nearly constant value at breakthrough. At a mobility of 0.1, the injectivity ratio then undergoes a decrease as complete areal sweep in the more permeable layer is approached.

Fig. 7.6 shows the volumetric sweep efficiency at breakthrough of a five-spot pattern, initially liquid filled, as a function of both mobility ratio and permeability variation.[2] This figure indicates that the major effect of mobility ratio on volumetric sweep efficiency at breakthrough occurs between mobility ratios of 0.1 and 10. Also as expected, the volumetric sweep at breakthrough decreases sharply as the coefficient of permeability variation increases. Figs. 7.7 and 7.8 show a similar plot for reservoirs having an initial gas saturation of 10 percent and 20 percent PV, respectively.

Note that Figs. 7.6 through 7.8 show only the breakthrough sweep efficiencies. Continued production after breakthrough would be expected to sweep ultimately all of the unswept area.

7.3 Influence of Gravity Forces

As early as 1933, Wyckoff et al.[4] recognized that because of its higher density, injected water would tend to move preferentially along the bottom of a formation.

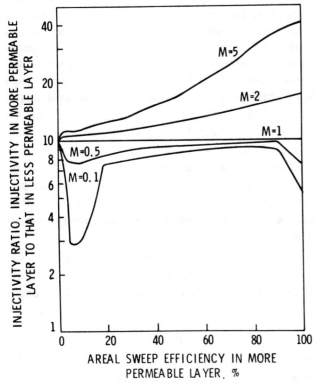

Fig. 7.5 Injectivity contrast in two-layered, liquid-filled five-spot.[2] 40-acre five-spot; 0.52-ft-radius wellbores; $d/r_w=1,780$; $k_1=100$ md; $k_2=10$ md.

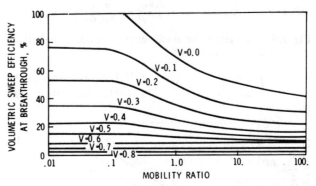

Fig. 7.7 Volumetric sweep efficiency at breakthrough, five-spot pattern; initial gas saturation=10 percent PV.[2]

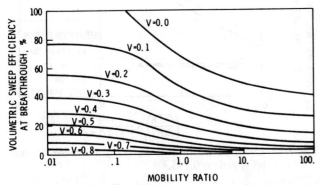

Fig. 7.6 Volumetric sweep efficiency at breakthrough, five-spot pattern; zero initial gas saturation.[2]

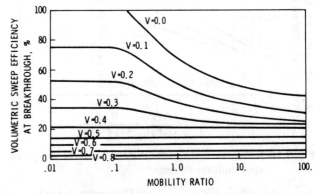

Fig. 7.8 Volumetric sweep efficiency at breakthrough, five-spot pattern; initial gas saturation=20 percent PV.[2]

This effect—gravity segregation of the injected fluid, resulting from the difference in density between the injected fluid and oil—was studied in detail by Craig et al.[5] Using miscible fluids, we studied the effect of gravity forces and mobility ratio with model tests that were dimensionally scaled.[6-8] Figs. 7.9 and 7.10 show the results for linear systems as well as for five-spot patterns. We found that the degree of gravity segregation of the injected fluid, measured in terms of volumetric sweep efficiency at breakthrough, depends upon the ratio of viscous forces to gravity forces, $\Delta P_h/\Delta P_v$. Higher rates resulted in higher volumetric sweeps. Over the expected range of this ratio in field operations, gravity effects in flat, uniform-permeability formations could result in oil recoveries at breakthrough as low as 20 percent of that otherwise expected. Devlikamov et al.[9] came to a similar conclusion.

Craig et al. studied also the gravity effects in a stratified five-spot-pattern model. The layers had a maximum permeability contrast of 50 to 1 and were in continuous flow communication so crossflow could occur. With the maximum permeability layer at either the top or bottom, very nearly the same results were obtained (Fig. 7.11). This suggests that, at least in some systems, the oil recovery at breakthrough can be affected to a greater degree by permeability stratification than by gravity forces.

7.4 Influence of Capillary Forces

In 1960 Gaucher and Lindley[10] reported the results of a waterflood study using a two-layer five-spot model in which viscous, capillary, and gravity forces were scaled simultaneously. The two layers were in flow communication throughout the interwell area. The sand was preferentially water-wet. The authors found that a marked difference in performance was obtained depending upon whether the more permeable layer was at the top or at the bottom, the higher sweeps at breakthrough being obtained when the more permeable layer was at the top of the model. The authors also observed that the volume of oil produced at a given injected water volume decreased slightly with rate whenever there was a tenfold change in water injection rate. Fig. 7.12 is a reproduction of Fig. 1 in their paper; it shows the performance of a simulated waterflood of a 2.17-cp oil from a reservoir having a layer permeability contrast of nearly 8 to 1.

In a continuation of Gaucher and Lindley's work, Carpenter et al.[11] simulated viscous, capillary, and gravity forces in a two-layered five-spot model. They deter-

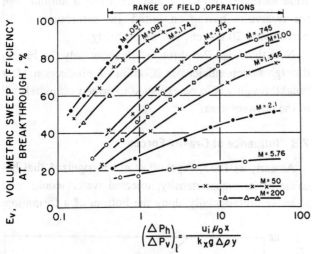

Fig. 7.9 Volumetric sweep efficiency at breakthrough, linear uniform systems.[5]

$$\left(\frac{\Delta P_h}{\Delta P_v}\right)_l = 2{,}050\,\frac{u_i(\text{B/D-sq ft})\,\mu_o\,(\text{cp})\,x\,(\text{ft})}{k_x\,(\text{md})\,\Delta\rho\,(\text{gm/cc})\,y\,(\text{ft})}$$

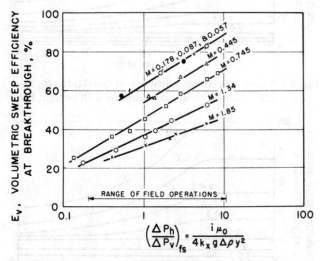

Fig. 7.10 Volmetric sweep efficiency at breakthrough, five-spot uniform systems.[5]

$$\left(\frac{\Delta P_h}{\Delta P_v}\right)_{fs} = 512\,\frac{i\,(\text{B/D-well})\,\mu_o\,(\text{cp})}{k_x\,(\text{md})\,\Delta\rho\,(\text{gm/cc})\,y^2\,(\text{sq ft})}$$

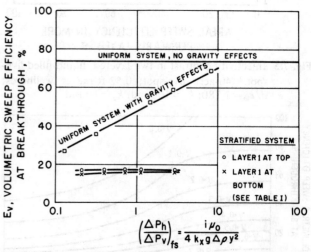

Fig. 7.11 Effect of nonuniformity on gravity effects; mobility ratio = 0.745.[5]

$$\left(\frac{\Delta P_h}{\Delta P_v}\right)_{fs} = 512\,\frac{i\,(\text{B/D-well})\,\mu_o\,(\text{cp})}{k_x\,(\text{md})\,\Delta\rho\,(\text{gm/cc})\,y^2\,(\text{sq ft})}$$

VERTICAL AND VOLUMETRIC SWEEP EFFICIENCIES

mined that the highest oil recovery was obtained at the lowest simulated water injection rate (range: 0.49 to 26.3 B/D/ft). They also found that the oil recovery performance obtained for the communicating layered system was between those observed for a homogeneous and for a noncommunicating layered system (Fig. 7.13).

The effect of viscous forces—i.e., rate—on vertical sweep efficiency is discussed in Section 7.6.

7.5 Crossflow Between Layers

The previously discussed studies by Craig et al.,[5] Gaucher and Lindley,[10] and Carpenter et al.[11] included consideration of the effects of crossflow between layers on the vertical sweep efficiency. Other studies have dealt specifically with that effect.

Hutchinson[12] presented some preliminary experimental results indicating that at favorable mobility ratios, the crossflow effects can increase the vertical coverage at breakthrough, and that at unfavorable mobility ratios the crossflow results in poorer sweep than is obtained with insulated layers (Fig. 7.14). Mathematical studies[13] arrived at the same conclusions as those obtained experimentally; that is, at very favorable mobility ratios, crossflow acts to increase the oil recovery efficiency, but at very unfavorable mobility ratios, the reverse is true. In another mathematical study[14] limited to a unit mobility ratio and no density difference, it was found that wellbore plugging of a high-permeability zone in contact with the remainder of the pay interval was not effective in preventing bypassing.

A numerical study of crossflow effects was made by Goddin et al.[15] Their study was concerned with the effects of viscous and capillary forces in a two-dimensional, two-layered preferentially water-wet system.

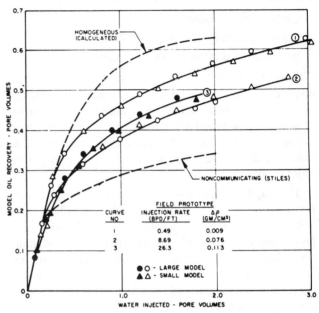

Fig. 7.13 Recovery histories for three pairs of five-spot experiments.[11]

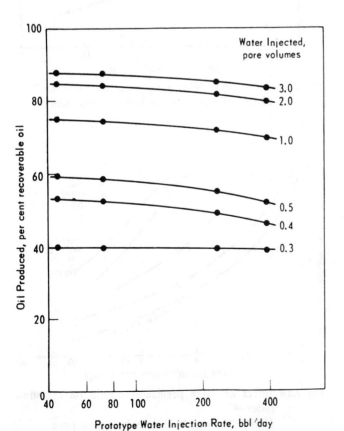

Fig. 7.12 Reservoir 2—oil recovery vs water-injection rate.[10]

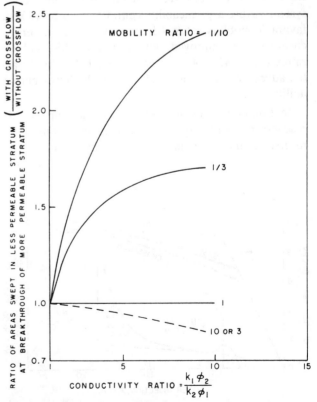

Fig. 7.14 Effect of cross-flow in a two-strata system as a function of conductivity ratio and mobility ratio. Results are preliminary.[12]

With mobility ratios from 0.21 to 0.95, they concluded that the computed oil recovery with crossflow was intermediate between that for a uniform reservoir and that for a layered reservoir with no crossflow (Fig. 7.15). This agrees with the experimental findings.[11] They defined a crossflow index this way:

$$\text{Crossflow index} = \frac{N_{pcf} - N_{pncf}}{N_{pu} - N_{pncf}} \quad \ldots \quad (7.4)$$

The additional subscripts to N_p are as follows:

- u — uniform system,
- cf — layered system with crossflow,
- ncf — layered system with no crossflow.

This term, computed at water breakthrough or at a given cumulative injected water volume, is a measure of how closely the performance of the system with crossflow approaches that of a homogenous system. At a crossflow index of zero, the performance is that of an insulated layer system; at one, that of a uniform permeability system.

The capillary:viscous-force ratio (inversely proportional to rate) had a minor effect on crossflow index (Fig. 7.16), but mobility ratio (Fig. 7.17) and layer permeability contrast (Fig. 7.18) had a marked influence. Improved recoveries at water breakthrough and beyond resulted from lower layer-permeability contrast, lower mobility ratios, and, to a lesser degree, from reduced flow rate.

In an experimental study of crossflow effects,[16] a series of miscible floods was made on a two-layered linear system having a permeability contrast of 20 to 1. With favorable mobility ratios (down to $M = 0.22$), vertical coverage was improved; but at unfavorable mobility ratios (up to $M = 4.7$), crossflow tended to hold back the advancement of the flood front in the lower permeability layer.

In summary, crossflow effects tend to improve the recovery performance at favorable mobility ratios; the reverse occurs with unfavorable mobility ratio.

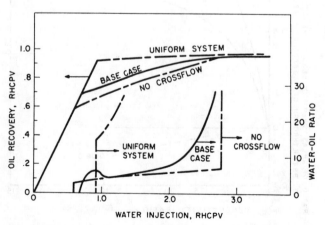

Fig. 7.15 Oil recovery and WOR for base case.[15] (RHCPV = recoverable hydrocarbon pore volume.)

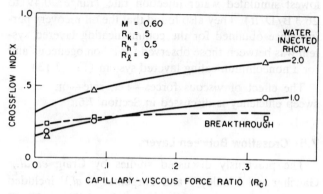

Fig. 7.16 Effect of ratio of capillary force to viscous force on crossflow index.[15] (RHCPV = recoverable hydrocarbon pore volume.)

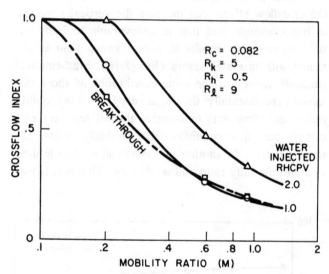

Fig. 7.17 Effect of mobility ratio on crossflow index.[15] (RHCPV = recoverable hydrocarbon pore volume.)

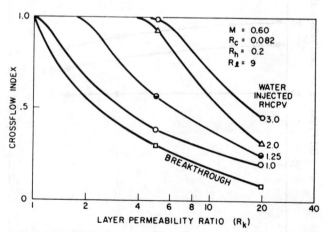

Fig. 7.18 Effect of layer permeability ratio on crossflow index.[15] (RHCPV = recoverable hydrocarbon pore volume.)

VERTICAL AND VOLUMETRIC SWEEP EFFICIENCIES

7.6 Effect of Rate on Volumetric Sweep Calculations

From 1957 to 1960 a controversy, fueled by technical papers and articles in the trade magazines,[17-23] raged on the rate sensitivity of waterflooding. The question: Is the oil recovery by waterflooding dependent upon either water-injection or oil-production rate? A second question then follows from an affirmative answer to the first: What is the optimum rate for maximum oil recovery? The discussion that follows will be limited to the effect of rate in water injection projects involving substantially flat horizontal reservoirs.

To explore the question of rate sensitivity of waterfloods, let us divide oil recovery into its two components: displacement efficiency and volumetric sweep efficiency.

Previous discussions (Chapter 3) have shown that the water-oil relative permeability characteristics of reservoir rock are independent of flow rate. Thus it follows that the oil displacement efficiency from reservoir rock is likewise independent of rate.

The two-dimensional term, vertical sweep efficiency, or its three-dimensional relative, volumetric sweep efficiency, is influenced, as we have seen, by viscous, capillary, and gravity forces. The viscous forces result from the pressure gradient and thus are proportional to the flow rate. In preferentially water-wet rock, the capillary forces cause the injected water to be imbibed into the smaller pores or less permeable lenses or strata within the reservoir. In rocks that are preferentially oil-wet, oil-water capillary forces tend to repel the injected water from smaller oil-filled pores. During the fillup stage of a waterflood, the less permeable zones in the reservoir become resaturated with oil as a result of the gas-oil capillary forces as well as the increased pressure in the oil bank. The gravity forces, dependent upon the density difference between the reservoir oil and water, act to pull the injected water to the lower portion of the reservoir.

In reservoirs having nonuniform permeability, the injected water moves preferentially through zones of higher permeabilty. In a preferentially water-wet rock, the capillary forces cause water to imbibe into adjacent, less permeable zones or lenses, while the ever-present gravity forces are acting to pull the injected water to the bottom of the reservoir.

In water-wet rocks, capillary forces can be efficient in displacing oil from less permeable portions of the reservoir. With lower water injection rates, more time is available for imbibition at and just behind the waterflood front. However, published information[10,11] suggests that rate variations of five or more have little effect on recovery.

The magnitude of imbibition depends upon three factors:

1. The range of rock permeabilites within the specific reservoir. The rock's attraction for the wetting phase is stronger the lower the permeability, but the ease of flow also decreases.
2. The size and location of the less permeable zones. This will control the number of opportunities for the injected water to imbibe into these streaks. Isolated, dispersed, tighter lenses will allow more imbibition and thus higher oil displacement than will occur if the low-permeability rock exists as a thick, continuous zone.
3. The wettability preference of the reservoir rock. The rate of imbibition will depend directly upon the degree of the rock's wettability preference for water.

The degree of gravity segregation depends upon the flow rate—the lower the water injection rate, the more severe the tendency for water to underrun the oil. Thus earlier water breakthrough is observed and a larger volume of injected water is required to produce the recoverable oil, resulting in higher producing WOR's. However, the degree of gravity segregation depends on the reservoir rock properties such as permeability and vertical transparency to fluid movement. Published information[5] indicates that significant changes in flow rate are required to effect small changes in the volumetric sweep efficiency resulting from gravity segregation.

To conclude, it is impossible to make a general statement as to optimum flooding rate because of the wide range of rock and fluid characteristics in oil reservoirs. The effect on each reservoir must be considered individually. In addition, published technical information suggests that injection rate changes of fivefold or more are required to significantly alter the effects of reservoir capillary or gravity forces.

7.7 Effect of Layer Selection on Volumetric Sweep Calculations

In preparing to use a waterflood prediction method, the reservoir engineer usually attempts first to define the vertical variation in permeabilities. The result is a description of the properties of a number of layers that constitute the reservoir. The engineer may choose at his discretion the number of layers to compose the formation. By selecting a large number of layers rather than a much smaller number, he knows that the most permeable layer will have a higher absolute permeability value, so that early water breakthrough into the production well can be expected. In addition, because the time and effort involved in performance predictions varies directly with the number of layers, these predictions may be costly. On the other hand, choosing a small number of layers can sometimes result in predicted performance approaching that of a uniform system.

A computer study[2] has provided some guidance in selecting the minimum number of layers. In this study the formation is considered to be composed of a number of layers, each of equal thickness and uniform permeability but with the layers' permeability having a log-normal probability distribution. The five-spot waterflood performance was calculated on the basis of a reservoir composed of one to 100 layers, with permeability variations ranging from 0.4 to 0.8. Although the calculated recovery at water breakthrough varied widely, depending upon the number of layers, the water and oil production performance at WOR's above about 3 varied less widely. Fig. 7.19 is a plot showing the effect of the number of layers upon the calculated waterflood performance at a permeability variation of 0.3 and a mobility ratio of 5.0. In some ranges of mobility ratio, the calculated after-breakthrough performance for as few as two layers agreed with that obtained using 100 layers. The waterflood performance obtained with 100 layers was arbitrarily chosen as a standard. The minimum number of equal-thickness layers that will yield calculated water and oil production performance equivalent to that for a 100-layer reservoir at WOR's above 2.5, 5.0, and 10.0, are given in Tables 7.1, 7.2, and 7.3, respectively. As both the mobility ratio and permeability variation increase, more layers are required to match the calculated 100-layer performance.

This discussion has dealt with equal-thickness, or equal-$h\phi$, layers. An alternative method of selecting layers is on the basis of equal flow capacity — i.e., equal kh. By such a method the more permeable layers would be thinner, the less permeable layers correspondingly thicker. Intuitively this seems to be a preferable technique for selecting layer properties, since the calculated waterflood performance would reflect in greater detail the performance of the more permeable layers. It is these layers that control the producing WOR and early breakthrough performance.

References

1. Muskat, M.: *Physical Principles of Oil Production*, McGraw-Hill Book Co., Inc., New York (1950) 682.
2. Craig, F. F., Jr.: "Effect of Permeability Variation and Mobility Ratio on Five-Spot Oil Recovery Performance Calculations", *J. Pet. Tech.* (Oct., 1970) 1239-1245.
3. Caudle, B. H. and Witte, M. D.: "Production Potential Changes During Sweep-Out in a Five-Spot Pattern", *Trans.*, AIME (1959) **216,** 446-448.
4. Wyckoff, R. D., Botset, H. G. and Muskat, M.: "The Mechanics of Porous Flow Applied to Water-Flooding Problems", *Trans.*, AIME (1933) **103,** 219-242.
5. Craig, F. F., Jr., Sanderlin, J. L., Moore, D. W. and Geffen, T. M.: "A Laboratory Study of Gravity Segre-

TABLE 7.1 — MINIMUM NUMBER OF EQUAL-THICKNESS LAYERS REQUIRED TO OBTAIN PERFORMANCE OF A 100-LAYER FIVE-SPOT WATERFLOOD AT PRODUCING WOR's ABOVE 2.5

(Confidence Level: Mean Square Difference ≤ 1 Percent Sweep)
(After Ref. 2)

Mobility Ratio	Permeability Variation							
	0.1	0.2	0.3	0.4	0.5	0.6	0.7	0.8
0.05	1	1	2	4	10	20	20	20
0.1	1	1	2	4	10	20	100	100
0.2	1	1	2	4	10	20	100	100
0.5	1	2	2	4	10	20	100	100
1.0	1	3	3	4	10	20	100	100
2.0	2	4	4	10	20	50	100	100
5.0	2	5	10	20	50	100	100	100

TABLE 7.2 — MINIMUM NUMBER OF EQUAL-THICKNESS LAYERS REQUIRED TO OBTAIN PERFORMANCE OF A 100-LAYER FIVE-SPOT WATERFLOOD AT PRODUCING WOR's ABOVE 5

(Confidence Level: Mean Square Difference ≤ 1 Percent Sweep)
(After Ref. 2)

Mobility Ratio	Permeability Variation							
	0.1	0.2	0.3	0.4	0.5	0.6	0.7	0.8
0.05	1	1	2	4	5	10	10	20
0.1	1	1	2	4	10	10	10	100
0.2	1	1	2	4	10	10	20	100
0.5	1	2	2	4	10	10	20	100
1.0	1	2	3	4	10	10	20	100
2.0	2	3	4	5	10	10	50	100
5.0	2	4	5	10	20	100	100	100

TABLE 7.3 — MINIMUM NUMBER OF EQUAL-THICKNESS LAYERS REQUIRED TO OBTAIN PERFORMANCE OF A 100-LAYER FIVE-SPOT WATERFLOOD AT PRODUCING WOR's ABOVE 10

(Confidence Level: Mean Square Difference ≤ 1 Percent Sweep)
(After Ref. 2)

Mobility Ratio	Permeability Variation							
	0.1	0.2	0.3	0.4	0.5	0.6	0.7	0.8
0.05	1	1	1	2	4	5	10	20
0.1	1	1	1	2	5	5	10	20
0.2	1	1	2	3	5	5	10	20
0.5	1	1	2	3	5	5	10	20
1.0	1	1	2	3	5	10	10	50
2.0	1	2	3	4	10	10	20	100
5.0	1	3	4	5	10	100	100	100

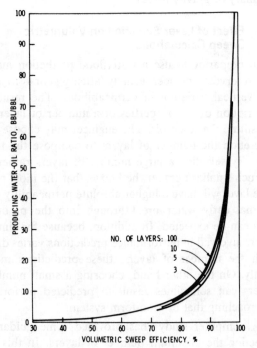

Fig. 7.19 Effect of number of layers on calculated waterflood sweep performance.[2] $V = 0.3$; $M = 5.0$.

gation in Frontal Drives", *Trans.*, AIME (1957) **210**, 275-282.

6. Geertsma, J., Croes, G. A. and Schwarz, N.: "Theory of Dimensionally Scaled Models of Petroleum Reservoirs", *Trans.*, AIME (1956) **207**, 118-127.

7. Rapoport, L. A.: "Scaling Laws for Use in Design and Operation of Water-Oil Flow Models", *Trans.*, AIME (1955) **204**, 143-150.

8. Perkins, F. M., Jr., and Collins, R. E.: "Scaling Laws for Laboratory Flow Models of Oil Reservoirs", *Trans.*, AIME (1960) **219**, 383-385.

9. Devlikamov, V. V., Sukhanov, G. N. and Bul'chuk, D. D.: "The Nature of Oil Displacement by Water Along the Thickness of a Formation", *Dzvest. Vysshikh. Ucheb Zaved.*, Neft i Gaz (1960) **3**, No. 1, 53-57.

10. Gaucher, D. H. and Lindley, D. C.: "Waterflood Performance in a Stratified Five-Spot Reservoir—A Scaled-Model Study", *Trans.*, AIME (1960) **219**, 208-215.

11. Carpenter, C. W., Jr., Bail, P. T. and Bobek, J. E.: "A Verification of Waterflood Scaling in Heterogeneous Communicating Flow Models", *Soc. Pet. Eng. J.* (March, 1962) 9-12.

12. Hutchinson, C. A., Jr.: "Reservoir Inhomogeneity Assessment and Control", *Pet. Eng.* (Sept., 1959) **31**, B19-26.

13. Warren, J. E. and Cosgrove, J. J.: "Prediction of Waterflood Performance in a Stratified System", *Soc. Pet. Eng. J.* (June, 1964) 149-157.

14. Root, P. J. and Skiba, F. F.: "Crossflow Effects During an Idealized Displacement Process in a Stratified Reservoir", *Soc. Pet. Eng. J.* (Sept., 1965) 229-238.

15. Goddin, C. S., Jr., Craig, F. F., Jr., Wilkes, J. O. and Tek, M. R.: "A Numerical Study of Waterflood Performance in a Stratified System With Crossflow", *J. Pet. Tech.* (June, 1966) 765-771.

16. Kereluk, M. J. and Crawford, P. B.: "Comparison Between Observed and Calculated Sweeps in Oil Displacement from Stratified Reservoirs", paper presented at API Mid-Continent District Meeting, Amarillo, Tex., April 3-5, 1968.

17. Funk, E. E.: "Effects of Production Restrictions on Water-Flood Recovery", *Prod. Monthly* (May, 1956) **21**, No. 5, 20-27.

18. Jordan, J. K., McCardell, W. M. and Hocott, C. R.: "Effect of Rate on Oil Recovery by Water Flooding", *Oil and Gas J.* (May 13, 1957) 99-108.

19. Richardson, J. G. and Perkins, F. M., Jr.: "A Laboratory Investigation of the Effect of Rate on Recovery of Oil by Water Flooding", *Trans.*, AIME (1957) **210**, 114-121.

20. Buckwalter, J. F., Edgerton, G. H., Stiles, W. E., Earlougher, R. C., Buckles, G. L. and Bridges, P. M.: "Waterflood Oil Recovery Is Lessened by Restricting Rates", *Oil and Gas J.* (June 15, 1958) 88-89.

21. Torrey, P. D.: "Effects of Curtailment on Waterflood Recovery", *Pet. Eng.* (Dec., 1958) **30**, B70-74.

22. McCardell, W. M.: "Further Discussion of 'Effects of Curtailment on Waterflood Recovery,' " *Pet. Eng.* (June, 1959) **31**, B128-133.

23. Menzie, D. E. and Cole, F. W.: "Waterflood Curtailment—Pro and Con", *Pet. Eng.* (May, 1960) **32**, B48-57.

Chapter 8

Methods of Predicting Waterflood Performance

A thorough study of waterflood prediction methods appeared in a series of articles published in 1968.[1] The same approach will be used here; that is, the perfect waterflood performance prediction method will be described and the currently available ones will be compared with that ideal. The waterflood prediction methods will be categorized in groups that consider *primarily*:

1. Reservoir heterogeneity
2. Areal sweep effects
3. Numerical methods
4. Empirical approaches

Where possible these prediction methods will first be described, then compared quantitatively; and comparisons will be made between predicted and actual waterflood performance. The practical use of these waterflood performance calculations will be discussed. In the last section of this chapter will appear recommended methods to obtain predictions of varying detail.

8.1 The Perfect Prediction Method

The perfect method for predicting waterflood performance would of course include all pertinent fluid flow, well pattern, and heterogeneity effects.

The fluid flow effects include the influence of different water-oil relative permeability characteristics as they differ from reservoir to reservoir as a result of wettability, pore size distribution, and connate saturations. The existence of a flood front (i.e., a zone of sharply increasing water saturation) would be included as well as consideration of any flowing oil behind the flood front, and the resulting change in fluid conductivity as the flood progresses. The possible presence of an initial gas saturation formed by solution gas drive depletion or gas injection prior to waterflooding would also be accounted for in the perfect prediction method.

Well pattern effects considered by the perfect prediction method would be the effect of mobility ratio on the areal sweep efficiency at water breakthrough as well as on the increase in areal sweep after breakthrough with continued water injection. The perfect prediction method would not be limited to a few patterns or injection-production well arrangements, but would be able also to predict the performance of peripheral floods and floods with irregularly placed injection wells.

Heterogeneity effects accounted for by the perfect program would include areal and vertical variations in permeability. Consideration of crossflow between adjacent segments of different permeability, as well as existence of any discrete isolated barriers to flow would be included. Of course the influence of viscous, capillary and gravity effects on the fluid movement would be considered by this prediction method.

Such a prediction method, since it models all the effects in waterflooding, would yield agreement between predicted and actual performance. But a perfect prediction method would require, too, detailed information on the reservoir's anatomy, probably more than we have today on any reservoir.

8.2 Prediction Methods Primarily Concerned with Reservoir Heterogeneity

Three basic types of prediction methods fall in this category:

1. Methods concerned with the effects of varying injectivity, layer by layer, in the radial portion of the reservoir surrounding the injection well.
2. Methods concerned with oil recovery, layer by layer.
3. Methods that characterize the reservoir nonuniformities by their permeability distribution and that calculate an over-all effect.

Yuster-Suder-Calhoun Method

In 1944, Yuster and Calhoun[2] developed equations approximating the variation in injectivity during a five-spot waterflood. They considered that the waterflood progressed through three stages: (1) radial outward movement of water from the injection well with a declining injectivity as the gas space becomes filled up, (2) an intervening period of water injectivity decline after interference from adjacent water injection wells until complete fillup, and (3) a final period of constant water injectivity. This approach was enlarged[3] to consider a reservoir whose heterogeneity could be simulated by a number of layers, each of different permeability, insulated from each other. It was assumed that the water and oil had equal mobilities and thus the portion of the injected water entering each layer was directly proportional to the fraction of the total flow capacity *(kh)* it represents. Piston-like displacement of the oil by water was assumed; that is, there was no flowing oil behind the flood front.

Muskat[4] extended the applicability of this method by considering the more general condition in which the water-oil mobility ratio can range from 0.1 to 10. He also discussed the effects of both linear and exponential permeability distributions.

Prats-Matthews-Jewett-Baker Method

Using basically the same approach, Prats *et al.*[5] proposed a more comprehensive method of predicting five-spot waterflood performance, including the combined effects of mobility ratio and areal sweep efficiency. The initial water injectivity is controlled by the mobilities of the injected water and oil banks. After water breakthrough a correlation is used that relates the injectivity with the radial portion of the producing well invaded by water. Piston-like displacement of oil by water is assumed. From any layer the production is either gas only (during the period of fillup), oil (during the period between fillup and water breakthrough), then water and oil, the proportion depending upon a laboratory-developed correlation of areal sweep and water cut. Given in Ref. 5 is an example of the application of their method.

The Prats *et al.* approach was applied by Fitch and Griffith[6] to the prediction of miscible flood performance.

Stiles Method

This method[7] basically involves accounting for the different flood-front positions in liquid-filled, linear layers having different permeabilities, each layer insulated from the others. Stiles assumes that the volume of water injected into each layer depends only upon the *kh* of that layer. This is equivalent to assuming a mobility ratio of unity. The Stiles method assumes that there is piston-like displacement of oil, so that after water breakthrough in a layer, only water is produced from that layer. After water breakthrough, the producing WOR is found as follows:

$$WOR = \frac{C}{1-C} \frac{k_{rw}}{\mu_w} \frac{\mu_o}{k_{ro}} B_o, \quad \ldots \quad (8.1)$$

where C is the fraction of the total flow capacity represented by layers having water breakthrough; thus producing water, and μ_w and μ_o are the water and oil viscosities, respectively. The Stiles method, therefore, contains an ambiguous condition regarding the oil and water mobilities—i.e., it assumes a unit mobility ratio in the vertical sweep calculations and takes into account in the calculation of producing WOR the mobility ratio that exists.

Schmalz and Rahme[8] presented the results of calculations using both the Stiles[7] and Suder and Calhoun[3] methods using six different permeability distributions. These distributions were categorized by the Lorenz coefficient (see Section 6.2). Other authors[9-11] allowed for layers of different properties, radial flow concepts, and oil bank buildup.

Johnson[12] developed a graphical approach that simplified the consideration of layer permeability and porosity variations. Layer properties were chosen such that each had equal flow capacities so that the volumetric injection rate into each layer was the same.

Dykstra-Parsons Method

An early paper[13] presented a correlation between waterflood recovery and both mobility ratio and permeability distribution. This correlation was based on calculations applied to a layered linear model with no crossflow. More than 200 flood pot tests were made on more than 40 California core samples in which initial fluid saturations, mobility ratios, producing WOR's, and fractional oil recoveries were measured. The permeability distribution was measured by the coefficient of permeability variation discussed in Section 6.2. The correlations presented by Dykstra and Parsons related the recovery at producing WOR's of 1, 5, 25, and 100 as a fraction of the oil initially in place to the permeability variation, mobility ratio, and the connate-water and floodout-water saturations. The values obtained presume a linear flood since they are based upon linear flow tests. An easy graphical technique[14] of using the Dykstra-Parsons method (Figs. 8.1 through 8.4) is available.

The Dykstra-Parsons technique was extended[15] to allow for liquid resaturation of gas space in each layer. Oil production from any layer then can occur only after fillup in that layer.

8.3 Prediction Methods Primarily Concerned with Areal Sweep

Muskat Method

In the 1940's, considerable work was done both mathematically and experimentally to determine the streamline and isopotential distributions in various flooding patterns.[16] These studies yielded the areal sweep efficiency at water breakthrough for a unit mobility ratio. Although this is not a waterflood prediction method as we know them today, operating engineers used these values in their estimates of waterflood recovery.

Hurst Method

Hurst[17] extended Muskat's earlier work for the five-spot pattern to consider the existence of an initial gas saturation prior to water injection. His mathematical studies considered the formation of an oil bank, but assumed the oil and water had equal mobilities. His was the first study to show the increase in areal sweep obtainable after water breakthrough by continued water injection.

Caudle et al. Method

Caudle and a series of coworkers[18-23] devoted a great deal of effort to the experimental studies of areal sweepout in a wide variety of flooding patterns. These patterns include the four-spot, five-spot, nine-spot, and line drive patterns. The work was extended[24] to the seven-spot pattern as well as the nine-spot. Using miscible fluids and the X-ray shadowgraph technique, they obtained values of four measures of performance: (1) areal sweep efficiency, (2) mobility ratio, (3) injected volume, and (4) portion of the production coming from the swept area. The injectivity variation during flooding was measured for many of these patterns. (See Section 5.6 for more details on these measures.) Because the studies were restricted to the use of miscible fluids, they apply for waterflood conditions in which there is no flowing oil behind the flood front. Other authors[25,26] have applied this basic technique to prediction of miscible fluid displacement projects.

Aronofsky Method

This method is based on potentiometric model studies[27,28] of the five-spot and line drive well arrange-

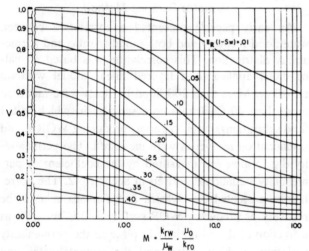

Fig. 8.1 Permeability variation plotted against mobility ratio, showing lines of constant $E_R (1 - S_w)$ for a producing WOR of 1.[14]

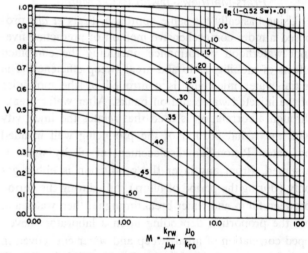

Fig. 8.3 Permeability variation plotted against mobility ratio, showing lines of constant $E_R (1 - 0.52 S_w)$ for a producing WOR of 25.[14]

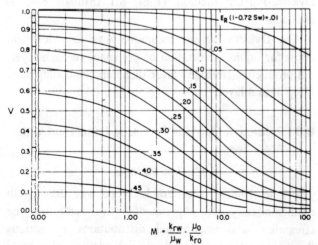

Fig. 8.2 Permeability variation plotted against mobility ratio, showing lines of constant $E_R (1 - 0.72 S_w)$ for a producing WOR of 5.[14]

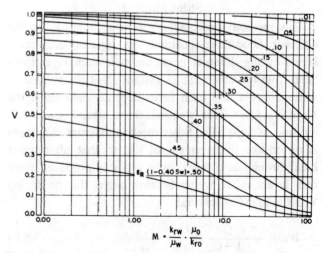

Fig. 8.4 Permeability variation plotted against mobility ratio, showing lines of constant $E_R (1 - 0.40 S_w)$ for a producing WOR of 100.[14]

ments. Areal sweep efficiencies at breakthrough were obtained as a function of mobility ratio for these two patterns. The variation in injectivity with areal sweep was determined for the five-spot. No sweep or injectivity data were presented beyond breakthrough, so this prediction method is limited to performance to water breakthrough. Piston displacement of the oil is presumed.

Deppe-Hauber Method

This method is based on two papers, the first by Deppe,[29] which presented information on the injectivity of pattern floods, and the second by Hauber,[30] who applied Deppe's results to calculations of pattern waterflood performance.

In Deppe's paper, the injectivity of a pattern flood is considered to be that of a series of linear and radial systems. In Hauber's paper, analytical expressions were derived for the five-spot and direct line drive patterns. For other patterns, it was presumed that the displacement took place along "stream tubes" that joined injection and production wells. Although an initial gas saturation could be handled, the method assumed that the oil saturation was instantly reduced to the residual oil saturation at the passage of the waterflood front. This method yielded good agreement with the breakthrough areal sweep data gathered on experimental systems.

8.4 Prediction Methods Dealing Primarily with Displacement Mechanism

The prediction methods to be discussed in this section are those concerned with frontal drives — that is, reflecting the possible presence of a saturation gradient and movable oil behind the waterflood front.

Buckley-Leverett Method

The Buckley-Leverett frontal advance theory,[31] as we have discussed before (Section 3.1), considers the mechanism of oil displacement by water in either a linear or a radial system. As an extension of this approach an equation was developed[32] for predicting the frontal advance rate in a radial system having an initial gas saturation. Welge's modification[33] to the frontal advance equation greatly simplified its use.

Other authors[34,35] combined the frontal drive equation with the Dykstra-Parsons prediction method to remove the previous limitation of piston-like displacement in each layer. In still another extension the Stiles method was modified[36] to consider Buckley and Leverett frontal drive effects.

Craig-Geffen-Morse Method

This prediction method[37] is based upon the results of a series of five-spot model gas and water drives. The approach is the use of a modified Welge equation and two experimentally derived correlations. The first correlation is that of areal sweep efficiency at breakthrough with mobility ratio. The second relates the areal sweep efficiency after breakthrough with the logarithm of the ratio W_i/W_{ibt}, where W_i is the cumulative injected water and W_{ibt} is that volume at water breakthrough. The second correlation can be expressed by the equation

$$E_A = E_{Abt} + 0.633 \log W_i/W_{ibt} \quad . \quad . \quad . \quad (8.2)$$

The method considers that the average water saturation in the water-contacted portion of the pattern is related to the cumulative injected water by a Welge-type equation modified to consider the "expanding water-contacted pore volume," caused by the increase in areal sweep. The oil production is taken to be the sum of the oil produced as a result of the increase in areal sweep and the oil displaced from the swept region. The water production is then the water injected minus the oil produced.

This method showed agreement with more than 20 laboratory model tests from which the correlations were obtained. These tests covered a range of mobility ratios and saturation gradients, and involved initial gas saturations up to 44 percent of the pore volume.

The method discussed in the original paper made no provision for multilayer reservoirs because for such conditions fluid injectivity correlations are needed to relate the injectivity into one layer with that into another layer. So that this calculation method can be applied to stratified reservoirs, the five-spot injectivity data of Caudle and Witte[20] have been used.

In a modification[38] termed the "Band Method", the reservoir is considered to comprise 10 bands of equal volume. Areal sweep information is incorporated, along with relative permeability effects, to yield the performance for any one layer. The assumption that the flow capacity of each layer does not change with time allows the performance of each band or zone to be added to yield composite performance. However, this assumption is equivalent to considering that the fluid injectivity performs as if at a unit mobility ratio. This method thus contains conflicting assumptions concerning the value of the mobility ratio.

In a 1969 paper[39] Wasson and Schrider discussed a method of predicting five-spot waterflood performance in stratified reservoirs. This method combined several previously published prediction techniques: that of Yuster and Calhoun[2] for calculating injection rate variation during the initial stages of fillup, that of Caudle and Witte[20] for determining the rate performance at fillup and beyond, and that of Craig et al.[37] for relating the injected water volume-produced oil-WOR performance. This prediction method in the form of a computer program is available for potential users. However a discussion of this paper,[40] pointed out that the effects of water-oil mobility ratio on performance were not fully considered.

Rapoport-Carpenter-Leas Method

In 1958, Rapoport et al.[41] proposed as a waterflood prediction method a laboratory-developed relationship between linear and five-spot flooding behavior. To predict the performance of a five-spot waterflood, a linear laboratory waterflood is performed on the sample of the reservoir in question (or its performance is calculated from water-oil relative permeability characteristics). The correlation relates the linear and five-spot recoveries at the same pore volumes of water injected through the oil-water viscosity ratio. There is no attempt to include areal sweep except as its effect is accounted for by the oil-water viscosity ratio. The authors established the correlation from flow tests on oil-wet glass beads, and implied that this same correlation would apply regardless of the wettability or of the porous medium.

Higgins and Leighton Method

In this method, frequently termed the streamline-channel flow method, a pattern flood can be considered to perform areally as a number of parallel flow tubes whose boundaries are the streamlines generated by a unit mobility ratio. First introduced in 1962[42] and followed by a number of other papers,[43-46] this method has shown itself to be extremely versatile. Although the water and oil at mobility ratios other than unity certainly do not follow precisely the flow channels existing at $M = 1$, the waterflood performance calculated by this method at favorable mobility ratios agrees with that calculated by other techniques.

For each stream tube or flow channel, shape factors that express the relative length and cross-sectional area are determined for equal-volume segments. The calculations for each flow channel are basically of the Buckley-Leverett type, with modifications to consider the varying cross-sectional area of each flow channel. The injectivity along any flow channel is computed from the saturation—and hence the relative permeability gradient—and the shape factors for each of its component cells. The performances of each of the flow channels (usually four in number) are summed to yield the composite performance.

A computer program for obtaining the required shape-factor data has been presented,[45] along with shape factors for the five-spot, seven-spot, and direct and staggered line drive pattern.

This streamline-channel flow technique has been applied[47] successfully to the prediction of peripheral waterflood performance. Shape-factor information for a variety of peripheral flood patterns has been discussed and published.[48-50]

8.5 Prediction Methods Involving Mathematical Models

As engineers became increasingly familiar with computers and sophisticated ways of solving complex mathematical problems, it was only a matter of time before mathematical models of waterfloods were developed.

Douglas-Blair-Wagner Method

One of the first papers dealing with a numerical analysis technique for both capillary and viscous effects was that of Douglas et al.[51] The reservoir system they simulated was linear, but it was the predecessor of a number of more complex mathematical models.

Hiatt Method

Hiatt[52] presented a detailed prediction method concerned with the vertical coverage or vertical sweep efficiency attained by a waterflood in a stratified reservoir. Using a Buckley-Leverett type of displacement, he considered, for the first time, crossflow between layers. The method is applicable to any mobility ratio, but is difficult to apply.

Douglas-Peaceman-Rachford Method

Extending the work of Douglas, Blair and Wagner in 1958, Douglas et al.[53] presented the results of a two-dimensional mathematical model that included the effects of relative permeabilities, fluid viscosities and densities, gravity, and capillary pressure. Thus it included all the necessary fluid flow effects and also considered two-dimensional well pattern effects. For practical use, however, this approach, with its completeness, requires a high-capacity, high-speed computer.

Warren and Cosgrove Method

Warren and Cosgrove[54] presented an extension of Hiatt's original work.[48] They considered both mobility ratio and crossflow effects in a reservoir whose permeabilities were log-normally distributed. No initial gas saturation was allowed, and piston-like displacement of oil by water was assumed. The displacement process in each layer is represented by a sharp "pseudointerface" as in the Dykstra-Parsons model.

Morel-Seytoux Method

This method[55,56] is primarily concerned with predicting the effect of pattern geometry and mobility ratio on waterflood recovery. Gravity and capillary effects are neglected; displacement is assumed to be piston-like and to occur at a unit mobility ratio. However, the results can be refined to account for two-phase flow and for mobility ratios other than unity. There are two steps to the approach: a numerical solution to obtain the pressure distribution at a unit mobility ratio, then an analytical technique to calculate injectivity, areal sweep at breakthrough, and subsequent producing WOR performance.

Other Mathematical Models

In addition to the previously discussed mathematical models, the more recent technical literature contains numerous references to prediction methods based upon the work of Douglas et al.[51,53] In these waterflood simulators, the reservoir is considered to be composed of a two-dimensional or three-dimensional network of rock, each segment of which can have different porosity, permeability, saturation, and relative permeability characteristics. With high-speed, high-capacity digital computers the perfect waterflood prediction method can be closely approached. These simulators can predict long-term reservoir performance in a few minutes of computing time. They are flexible enough to consider complex injection and production scheduling; and versions of them are widely used by both oil producers and consultants. However, as one reservoir engineer has put it: "You shouldn't use a sledge hammer to drive a tack." That is, for simple problems, use a simple tool; for more complex problems, a more sophisticated tool.

8.6 Empirical Waterflood Prediction Methods

Guthrie-Greenberger Method

In a 1955 paper by Guthrie and Greenberger,[57] oil recovery by water drive was related empirically to reservoir rock and fluid properties. They studied 73 sandstone reservoirs that had water drive or that had solution gas drive combined with water drive. Actual production data were available for these reservoirs. The oil recovery was related to the permeability, porosity, oil viscosity, formation thickness, connate water saturation, depth, oil reservoir volume factor, area, and well spacing. The correlation shown below fit so well that 50 percent of the time the recovery factor was within 6.2 percent of the reported value, and 75 percent of the time it was within 9.0 percent.

$$E_R = 0.2719 \log k + 0.25569 S_w - 0.1355 \log \mu_o \\ - 1.5380 \phi - 0.0003488 h + 0.11403, \quad (8.3)$$

where E_R is the fractional recovery efficiency.

This equation implies that water drive recovery efficiency is lower in reservoirs of higher porosity!

Schauer Method

Schauer[58] presented an empirical method for predicting the waterflood behavior of Illinois Basin waterfloods. This method is based on the past performance of five floods. A plot was constructed showing percentage fillup at first signs of an oil production response as a function of Lorenz coefficient. As Lorenz coefficient increased—that is, with reservoirs of increased non-uniformity—the oil production response, or "buzz", occurred at a lower percentage of fillup (Fig. 8.5). Other plots showing the injectivity decline with time were also obtained from field performance history.

Guerrero-Earlougher Method

Guerrero and Earlougher[59] presented a number of rules of thumb for predicting waterflood performance.

1. Oil production first begins when the injected water volume is from 60 to 80 percent of the gas-filled reservoir space.
2. Waterflood oil producing rates peak immediately after fillup and remain at that level for 4 to 10 months.
3. The period of peak production will occur when the ratio of the water injection rate to oil producing rate ranges from 2 to 12, with values of 4 to 6 considered average for a typical flood.
4. The oil production rate will decline thereafter at a rate of 30 to 70 percent/year.

These rules of thumb have limited applicability.

API Statistical Study

The API Subcommittee on Recovery Efficiency, headed by J. J. Arps, presented a statistical study of recovery efficiency.[60] From a statistical analysis of data from 312 reservoirs, they developed correlations for water drive recovery from sandstone and sand reservoirs and for solution gas drive recoveries from sandstones, sands, and carbonates. The water drive recovery, as a percent of the original oil in place, is:

$$E_R = 54.898 \left[\frac{\phi (1 - S_w)}{B_{oi}} \right]^{0.0422} \left[\frac{k \mu_{wi}}{\mu_{oi}} \right]^{0.0770}$$

$$(S_w)^{-0.1903} \left(\frac{p_i}{p_a} \right)^{-0.2159} \quad \ldots \quad (8.4)$$

This correlation for waterflood recovery expressed as a logarithmic-type equation depends upon porosity, connate water saturation, permeability, oil and water viscosities, initial pressure (p_i) and pressure at depletion (p_a). The correlation coefficient for the equation is 0.958, which by its closeness to 1.000 shows a very good fit of the data. This correlation developed from *water drive* reservoir performance data has limited usefulness for *waterflooding* projects.

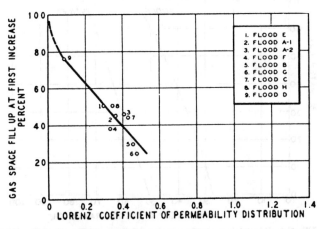

Fig. 8.5 Correlation of Lorenz coefficient with gas volume filled up at first production increase.[58]

Other Correlations

Correlations for estimating waterflood performance have been developed from histories of floods in Oklahoma[61] and the Denver Basin in Colorado and Nebraska.[62] The usefulness of this sort of correlation is generally limited to reservoirs in the particular geographical area being studied.

8.7 Comparisons of Performance Prediction Methods

Table 8.1 compares the various waterflood prediction methods with the perfect method. Shown are the methods and a check list showing whether they consider various facets of fluid flow, pattern, and heterogeneity effects. A quick glance will reveal that none of the listed prediction methods developed to date, except the more recent mathematical models, satisfy the requirements of the perfect method. In the following portions of this section, we shall see how the various waterflood prediction methods compare.

Stiles and Yuster-Suder-Calhoun Methods

The first comparison of prediction methods was shown by Schmalz and Rahme.[8] Different permeability distributions were used and waterflood performance was calculated by the Stiles method and by the Yuster-Suder-Calhoun (Suder) method. These comparisons are shown as Figs. 8.6, 8.7, and 8.8. The figures are based on a straight-line permeability distribution, a probability (normal) distribution, and an actual field permeability distribution, respectively. Fig. 8.9 shows a comparison of both these prediction methods correlated with Lorenz coefficient. There is wide divergence in predicted performance, which is to be expected since the Suder method assumes a log-normal permeability distribution, and the Stiles method has no such restriction.

Dykstra-Parsons, Stiles, Suder-Calhoun, and Felsenthal et al. Methods

In 1962, Felsenthal et al.[15] compared the performance predicted by their method with that predicted by Dykstra-Parsons, Stiles, and Suder-Calhoun. Fig. 8.10 evaluates the optimum pressure at which a waterflood

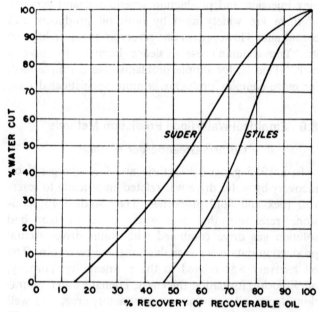

Fig. 8.7 Water cut vs recovery — Stiles and Suder methods for the probability distribution.[8]

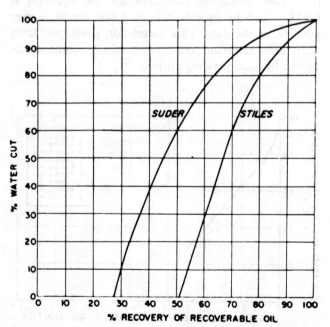

Fig. 8.6 Water cut vs recovery — Stiles and Suder methods for the straight-line distribution.[8]

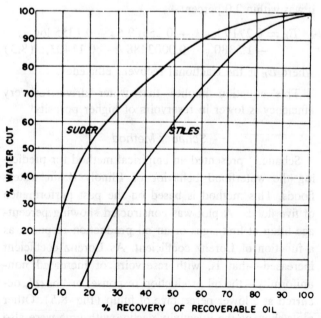

Fig. 8.8 Water cut vs recovery — Stiles and Suder methods for factual field data.[8]

PREDICTING WATERFLOOD PERFORMANCE

TABLE 8.1 — COMPARISON OF WATERFLOOD PREDICTION METHODS

Method and Modification	Date Presented	Fluid Flow Effects			Applies to			Pattern Effects					Heterogeneity Effects		
		Consider Initial Gas Saturation?	Consider Saturation Gradient?	Consider Varying Injectivity?	Linear System?	Five-spot Pattern?	Other Patterns?	Applicable Mobility Ratio?	Consider Areal Sweep?	Consider Increase Sweep After Breakthrough?	Require Published Lab Data?	Require Additional Lab Data?	Consider Stratified Reservoir?	Consider Cross-flow?	Consider Spatial Variations?
Perfect Method		yes	yes	yes	yes	yes	yes	any	yes	yes	no	no	yes	yes	yes
1. Yuster-Suder-Calhoun	1944	yes	no	yes	—	yes	no	1.0	no	no	no	no	yes	no	no
Muskat	1950	no	no	yes	yes	no	no	any	no	no	no	no	yes	no	no
Prats et al.	1959	yes	no	yes	—	yes	no	any	yes	yes	yes	yes	yes	no	no
Dykstra-Parsons	1950	yes	no	no	yes	no	no	any	no	no	no	no	yes	no	no
Johnson	1956	yes	no	no	yes	no	no	any	no	no	no	no	yes	no	no
Felsenthal et al.	1962	yes	no	no	yes	no	no	any	no	no	no	no	yes	no	no
Stiles	1949	no	no	no	yes	no	no	1.0	no	no	no	no	yes	no	no
Schmalz-Rahme	1950	no	no	no	yes	no	no	1.0	no	no	yes	no	yes	no	no
Arps	1956	no	no	no	—	yes	no	1.0	no	no	no	no	yes	no	no
Ache	1957	no	no	no	yes	yes	no	1.0	no	no	no	no	yes	no	no
Slider	1961	yes	no	yes	yes	yes	no	1.0	no	no	no	no	yes	no	no
Johnson	1965	yes	no	yes	yes	no	no	1.0	no	no	no	no	yes	no	no
2. Muskat	1946	no	no	no	no	yes	yes	1.0	yes	no	no	no	no	no	no
Hurst	1953	no	no	no	—	yes	no	1.0	yes	yes	no	no	yes	no	no
Caudle et al.	1952-59	no	no	yes	—	yes	yes	any	yes	yes	yes	no	yes	no	no
Aronofsky	1952-56	no	no	yes	—	yes	no	any	yes	no	yes	no	yes	no	no
Deppe-Hauber	1961-64	yes	no	no	—	yes	yes	1.0	yes	no	yes	no	yes	no	no
3. Buckley-Leverett	1942	no	yes	no	yes	no	no	—	no	no	no	no	no	no	no
Roberts	1959	no	yes	no	yes	no	no	—	no	no	yes	no	yes	no	no
Kufus and Lynch	1959	no	yes	no	yes	no	no	any	no	no	no	no	yes	no	no
Snyder and Ramey	1967	no	yes	no	—	no	no	any	no	no	no	no	yes*	no	no
Craig-Geffen-Morse	1955	yes	yes	yes*	—	yes	no	any	yes	yes	yes	no	no	no	no
Hendrickson	1961	no	yes	no	—	yes	no	any	yes	yes	yes	no	yes	no	no
Wasson and Schrider	1968	yes	yes	yes*	—	yes	no	any	yes	yes	yes	yes	yes*	no	no
Rapoport et al.	1958	no	yes	no	yes	yes	no	any	yes	yes	yes	no	yes	no	no
Higgins-Leighton	1962-64	yes	yes	yes	yes	yes	yes	any	yes	yes	yes	no	yes	no	no
4. Douglas-Blair-Wagner	1958	no	yes	no	—	yes	yes	any	yes	yes	yes	no	yes	yes	no
Hiatt	1958	no	no	no	—	no	no	any	no	yes	no	no	yes	yes	no
Douglas et al.	1959	no	yes	no	—	yes	yes	—	yes	yes	no	no	yes	yes	no
Warren-Cosgrove	1964	yes	yes	no	yes	no	no	any	no	no	no	no	yes	no	no
Morel-Seytoux	1965-66	no	yes	yes	—	yes	no	any	yes	yes	no	no	yes	no	no
5. Guthrie-Greenberger	1955	no	yes	no	—	—	yes	any	yes	yes	no	no	yes	yes	yes
Schauer	1957	yes	no	no	—	yes	no	any	yes	no	no	no	yes	no	no
Guerrero-Earlougher	1961	yes	no	no	—	yes	yes	—	no	no	no	no	yes	no	no
API	1967	no	yes	no	—	—	yes	any	yes	yes	no	no	yes	yes	yes

*Using Caudle and Witte[20] injectivity correlation

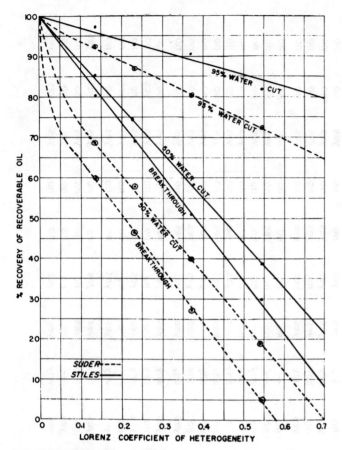

Fig. 8.9 Comparison between Suder and Stiles waterflood predictions.[8]

should be started to obtain the maximum total recovery from solution gas drive and waterflood. The reservoir considered was very nonuniform, having a coefficient of permeability variation of 0.9. At any pressure level, the Stiles method indicated the highest recovery and the Felsenthal et al. "resaturation method" indicated the lowest. Fig. 8.11 shows the results of a similar study for a permeability variation of 0.6, a less stratified reservoir. Fig. 8.12 compares the Stiles and Dykstra-Parsons methods for different mobility ratios at a permeability variation of 0.6. Over the range of water-oil mobility ratios, both methods yield about the same recovery at a producing WOR of 25.

Craig et al. and Higgins-Leighton Methods

A series of comparisons was made of single-layer waterflood performance by the Craig et al. and Higgins-Leighton methods. This comparison covered a range of mobility ratios from 0.2 to 2.1. Fig. 8.13 shows a comparison made at a mobility ratio of 0.2. In general, the prediction methods gave equivalent results, with close agreement particularly at the more favorable mobility ratios.

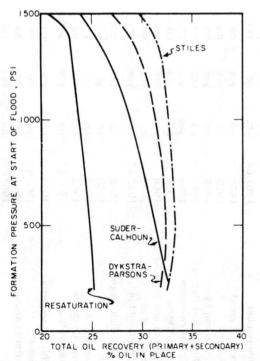

Fig. 8.10 Total recovery as related to the reservoir pressure at start of waterflood.[15] Permeability variation = 0.9.

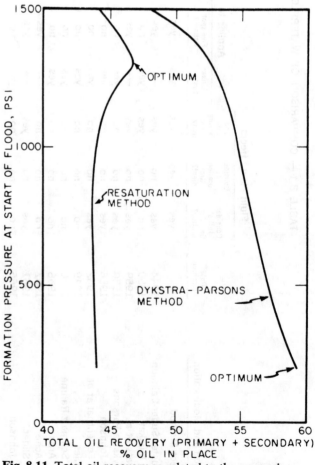

Fig. 8.11 Total oil recovery as related to the reservoir pressure at start of waterflood.[15] Permeability variation = 0.6.

Stiles, Dykstra-Parsons, and Snyder-Ramey Methods

In 1967, Snyder and Ramey[36] presented a comparison of the Stiles, Dykstra-Parsons, and Snyder-Ramey methods using a 10-layer linear system. The layer properties had a log-normal permeability distribution with a variation of 0.5. Fig. 8.14 compares the calculated results for a mobility ratio of 0.125. Two Buckley-Leverett models were considered: Model No. 1, in which all properties vary between layers; and Model No. 2, in which only permeability varies between layers, and porosity, connate water saturation and residual oil saturation are the same in each stratum. As compared with the Snyder-Ramey method, both the Stiles and Dykstra-Parsons methods gave low values for breakthrough recovery and pessimistic predictions for performance after breakthrough. Fig. 8.15 shows a similar comparison for a mobility ratio of 1.0. Both Stiles and Dykstra-Parsons methods gave the same results at this condition ($M = 1$, and permeability distribution is log normal); and the results were very close to those of Model No. 2 (the model in which only the permeability varies between layers).

8.8 Comparisons of Actual and Predicted Performance

In their comparison of waterflood prediction methods, Guerrero and Earlougher[59] showed the actual and the predicted performance for two floods. This comparison is shown in Figs. 8.16 through 8.19. A wide divergence in predicted performance is shown, with the major differences being in the predicted peak producing rates and the cumulative oil recoveries.

The most quoted waterflood data are those originally reported by Prats et al.,[5] then quoted by Slider[11] in support of his own method, and finally compared by Higgins and Leighton[43] with their performance predictions (Fig. 8.20). Higgins and Leighton obtained a better fit than had either Slider or Prats et al.

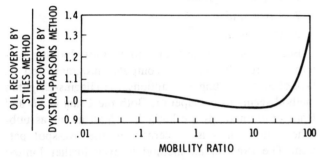

Fig. 8.12 Comparison of calculated oil recovery at a WOR of 25 — Stiles and Dykstra-Parsons methods.[15] Permeability variation (V) is 0.6.

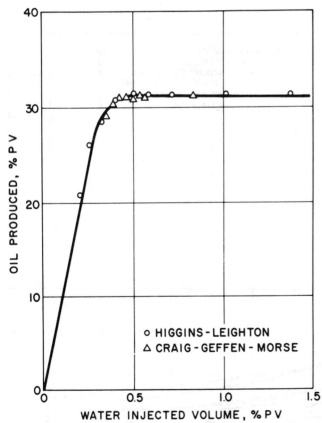

Fig. 8.13 Comparison between Higgins-Leighton and Craig et al. five-spot waterflood performance calculations. Mobility ratio = 0.2.

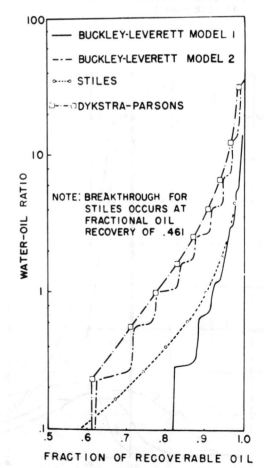

Fig. 8.14 Comparison of Stiles, Dykstra-Parsons and Snyder-Ramey waterflood predictions.[36] $M = 0.125$; $V = 0.5$.

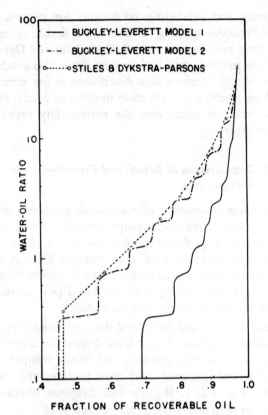

Fig. 8.15 Comparison of Stiles, Dykstra-Parsons, and Snyder-Ramey waterflood predictions.[36] $M = 1.0$; $V = 0.5$.

In 1964, Abernathy[63] compared the observed five-spot pilot performance of three West Texas carbonate reservoirs with that predicted by the Stiles,[7] Craig et al.,[37] and Hendrickson[38] methods. The pilot tests were conducted in the Panhandle, Foster and Welch fields. Figs. 8.21 through 8.23 show the performance of these three floods. The Craig et al. method was found to closely approximate the water-cut vs recovery performance of the three floods but was less than completely satisfactory in predicting water injection rates. This method was regarded by Abernathy as superior to both the Stiles and Hendrickson approaches.

Of the methods discussed, three appear most promising. These are in order:

1. Higgins-Leighton
2. Craig et al.
3. Prats et al.

Because of its adaptability to various flooding patterns, its readily available computer program, and its lack of many limiting assumptions, the Higgins-Leighton method seems to be superior. Both the Craig et al. and Prats et al. methods are limited, at least as far as published information is concerned, to the five-spot pattern. The Prats et al. method is even further limited since the laboratory-developed data necessary for applying it to mobility ratios other than that discussed in the paper are not available. All of these methods are limited to no-crossflow conditions.

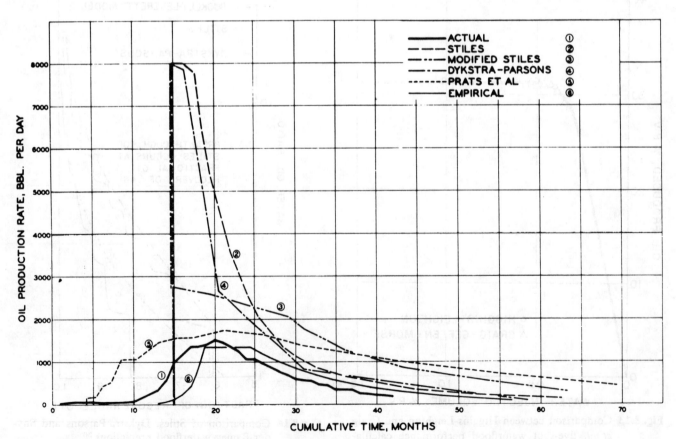

Fig. 8.16 Comparison of actual and predicted recovery history, Flood 1.[59]

PREDICTING WATERFLOOD PERFORMANCE

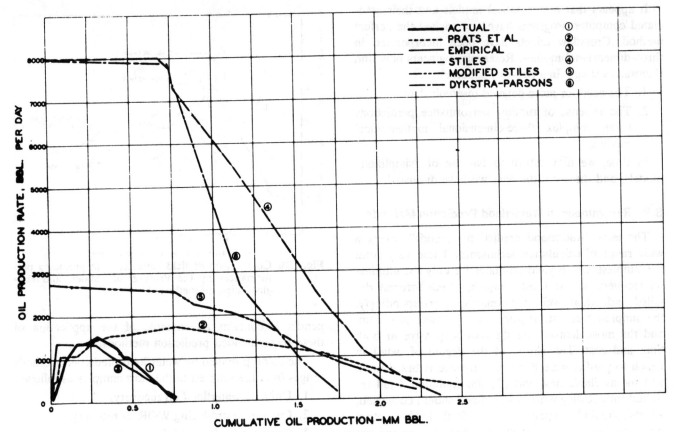

Fig. 8.17 Comparison of actual and predicted oil recovery, Flood 1.[59]

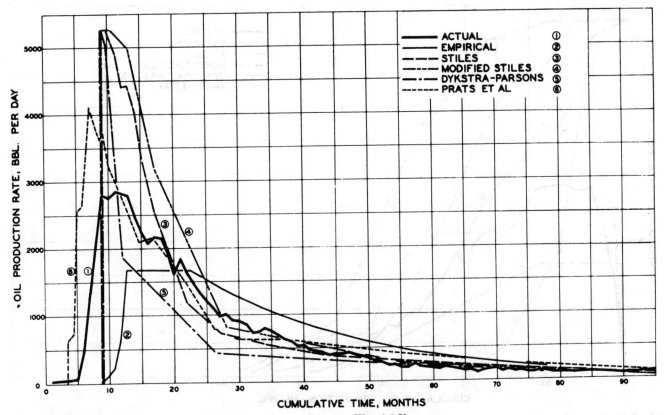

Fig. 8.18 Comparison of actual and predicted recovery history, Flood 2.[59]

It appears that mathematical models and their associated computer programs have approached the perfect method. Crossflow effects have been incorporated in three-dimensional models. Reservoir engineers now find themselves caught in a two-way squeeze:

1. The need for more detailed reservoir data.
2. The expense of running performance predictions using complex three-dimensional mathematical models.

In time, we may return to the use of "simplified" models, and these are the ones we have discussed.

8.9 Recommended Waterflood Prediction Methods

The term "waterflood prediction methods" covers a wide range of calculation techniques. These vary from the simplest, which yield estimates for only the ultimate oil recovery, to the most complex, which forecast detailed individual well performance. Correspondingly, the simplest forecasts require the least time to obtain, and the most detailed are the most expensive in both time and cost. The choice of the degree of detail in which to predict waterflood performance is based upon (1) the available time and (2) the minimum detail required for deciding when a waterflood should be started, when it should be expanded, and so forth. In this section we shall discuss the ranges of detail that waterflood prediction methods provide and the recommended methods for obtaining various degrees of detail. Ap-

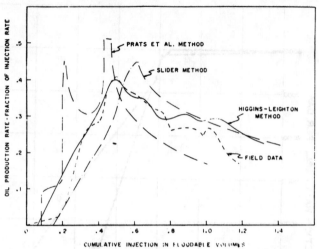

Fig. 8.20 Comparison of field behavior with predicted performances by the Prats *et al.*, Slider, and Higgins-Leighton methods.[43]

pendix E contains an example of the application of these recommended prediction methods.

The detail provided by waterflood prediction methods ranges from the simplest to the most complex as follows:

1. Ultimate waterflood oil recovery.
2. Composite producing WOR vs recovery.
3. Composite values of injection rate, producing rate, producing WOR, oil recovery, and cumulative injected water, all vs time.

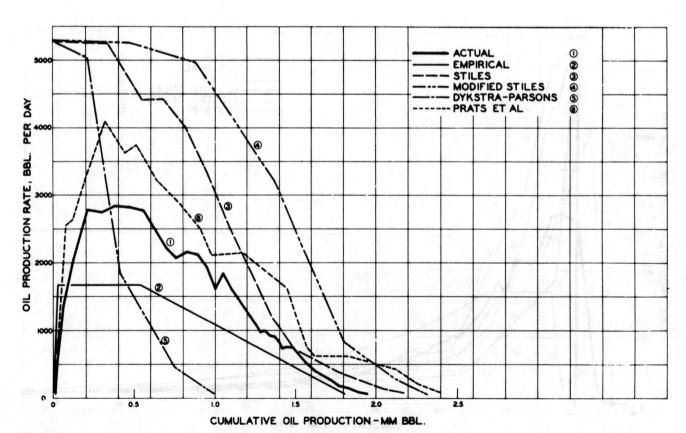

Fig. 8.19 Comparison of actual and predicted oil recovery, Flood 2.[59]

4. Individual injection rates and cumulatives and individual producing rates, water-oil ratios, and oil recoveries, all vs time.

It should be recognized that more detailed predictions require more detailed reservoir data.

Ultimate Waterflood Recovery

The simplest estimate to obtain is that of ultimate waterflood recovery. This generally can be done in a few hours after the required data are assembled. The necessary information consists of the average water-oil relative permeability characteristics applying to the reservoir (see Section 2.6), the reservoir oil and water viscosities, the initial water saturation, and the oil reservoir volume factor at the pressure corresponding to the start of waterflooding.

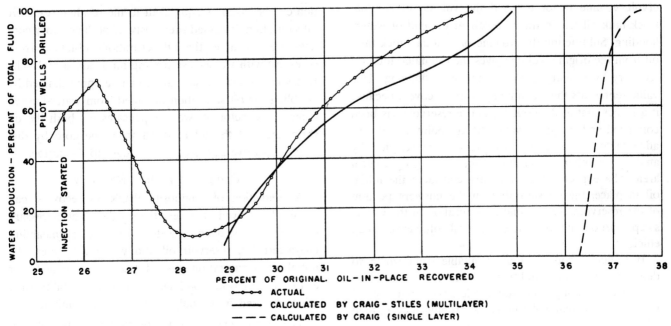

Fig. 8.21 Calculated vs actual performance of Panhandle field.[63]

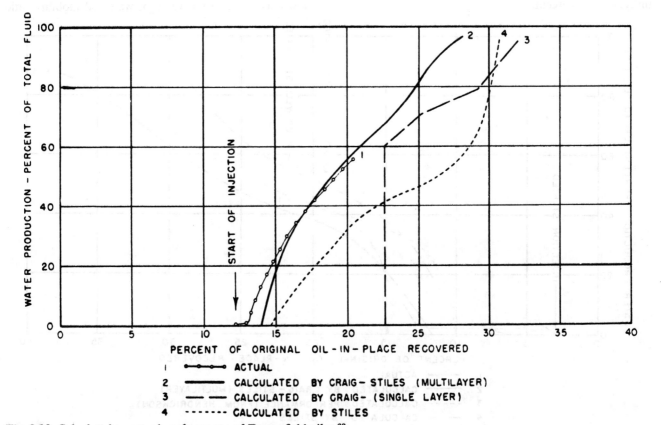

Fig. 8.22 Calculated vs actual performance of Foster field pilot.[63]

Using the relative permeability data, a fractional flow curve is drawn. From this fractional flow curve the average water saturation corresponding to the economic producing water-cut limit is determined. It is usually acceptable to assume that there is no free gas remaining in the reservoir at floodout. We can also usually assume that solution of the free gas causes no significant change in either oil viscosity or oil reservoir volume factor. The volume of residual reservoir oil is reduced by the oil reservoir volume factor to yield the volume of residual stock-tank oil left in the reservoir at the end of waterflooding. Subtracting this volume of residual stock-tank oil from the original volume gives the estimated total oil recovery by primary depletion and waterflooding. This value represents the maximum possible recovery because it assumes that at floodout the entire reservoir has been contacted and swept by water. At this point a conformance factor, or sweep efficiency, can be selected by analogy to other reservoirs in the same geographical area. The expected total recovery considers the initial oil in place, the oil saturation in the unswept portion of the reservoir, the residual oil saturation in the water-swept portion and, of course, the total volumetric sweep efficiency.

If sufficient core data are available from which to determine the permeability variation, V, a better approach to selecting total volumetric sweep efficiency is possible. The total volumetric sweep efficiency is then $\frac{(1-V^2)}{M}$, where the maximum value of this fraction is unity, or 100 percent.

Subtracting the primary oil recovery from the expected oil recovery will yield the expected waterflood recovery. This can be expressed in barrels, barrels per acre, barrels per acre-foot, or percent of original oil in place.

Frequently, water-oil relative permeability data are not available, the only information on oil displacement efficiency being end-point residual oil data from core tests similar to flood pot tests. This residual oil saturation can be taken as equivalent to the residual reservoir oil saturation discussed previously. If no laboratory core data are available, the oil saturations obtained from cores cut with water-base drilling muds are useful. The reservoir residual oil saturation can be approximated by doubling the oil saturation obtained from water-flushed cores. This factor of two compensates for both the oil shrinkage and the oil expelled from the core by dissolved gas drive as the core is brought to the surface.

Composite WOR vs Recovery

A prediction of composite WOR vs recovery can usually be arrived at in less than a day. The necessary data include the average water-oil relative permeability characteristics, reservoir oil and water viscosities, the initial water saturation, the oil reservoir volume factor at the start of waterflooding, and the permeability variation obtained from analysis of core permeabilities.

The recommended calculation procedure is that proposed by Dykstra and Parsons.[13] Using the fractional flow curve, obtain the value of water-oil mobility ratio.

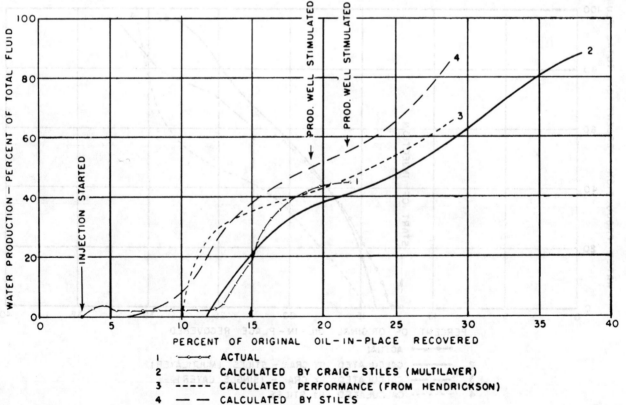

Fig. 8.23 Calculated vs actual performance of Welch field pilot.[63]

From the mobility ratio, the permeability variation, and the connate water saturation, obtain the fractional oil recovery, R, at producing water-oil ratios of 1, 5, 25, and 100, using published correlation charts[14] (Figs. 8.1 through 8.4). This fractional recovery is the portion of the original oil in place recovered by both primary methods and waterflooding. The limitation of these correlation charts is that they inherently assume that the residual oil saturation by waterflooding is the same as that in California sands.

Composite Injection and Producing Rates, WOR and Recovery vs Time

Composite WOR and recovery vs time can be predicted in a few days by hand calculation or in a few minutes by computer. The necessary data are those required for composite WOR vs recovery.

The recommended calculation procedure is that for the oil recovery performance proposed by Craig et al.,[37] together with Muskat's injectivity relationship[64] for the fillup period and Caudle and Witte's injectivity correlation for liquid-filled systems.[20] Although all of these papers deal with the five-spot well pattern exclusively, the oil recovery performance of any repeating pattern can be approximated by that for a five-spot. As an alternative, the Higgins-Leighton method[40] may be used, but it requires a computer.

In their paper Craig et al. present a tabular worksheet for the calculation of oil recovery and producing WOR as a function of cumulative injected water volume. The injected water is related to time during the fillup stage of the flood by the radial injectivity equation (Eq. 7.3), and after fillup by the injectivity correlation of Caudle and Witte.

From the permeability variation and the mobility ratio, the minimum number of layers that must be considered is obtained from Table 7.1, 7.2, or 7.3, depending upon the extent of the WOR performance for which adequate performance is required. Generally the number of layers obtained from Table 7.1 are few enough that the calculations are not too time-consuming.

The waterflood performance of each layer is calculated and all of them are combined at corresponding times to yield total (i.e. multilayer) performance. Where the properties of each layer are the same except for permeability, the performance of only one layer need be calculated in detail. The performance of each remaining layer can be obtained by "sliding the time scale" as shown in Appendix E. That is, if the second layer, for example, is only half as permeable as the first, it will require twice the time to reach a given injected water volume and corresponding oil recovery and WOR. At that injected water volume the injection and producing rates for the second layer will be half those for the first layer.

If the proposed waterflood is of the peripheral or end-to-end type involving several rows of producers off-setting the injection wells rather than a pattern flood, the approach outlined above is inadequate. For these situations no simple approach is really satisfactory. Part of the reason for this is that the distances from injectors to producers are no longer equal or nearly so. The actual producing operation is much more complex since decisions on shutting in producers or moving injectors must be made from time to time and cannot be forecast with any assurance. Qualitatively, however, it may be said that the waterflood performance calculated by the foregoing method will yield: (1) acceptable initial per-well injection rates; and (2) pessimistic oil-recovery vs WOR performance.

Individual Well Performance

To predict individual well performance during a waterflood, calculations frequently can require a month or more, and in some cases many months. Such a prediction requires a computer program of the type discussed previously.[53] Depending upon the mathematical model (computer program) used, single-layer or multi-layer (with or without crossflow) performance can be calculated.

These computer programs generally require a detailed description of the reservoir. Reservoir properties at each of a multitude of grid points must be specified. As a result, data preparation for these mathematical models is time-consuming. Since these properties can exert a large influence on the results, they must be carefully chosen.

The more complex mathematical models can take into account the effect on performance of producing and injection well stimulation some time in the future, the shutting in of producing wells as a specified water-cut is exceeded, the conversion of producers to injectors, restrictions on allowable producing rate, and many other factors related to month-to-month and year-to-year operation.

These models can be extremely versatile, but at the same time they can be quite expensive. Depending upon how complex the models are, reservoir studies can consume hundreds to thousands of dollars in computer cost.

8.10 Practical Use of Waterflood Prediction Methods

The practical use of waterflood prediction methods is of course to forecast future oil production performance. To use a prediction method for a reservoir about to be waterflooded, one must be able to specify the water-oil flow properties, the initial fluid saturations, and most importantly a description of the reservoir and its permeability variation, both laterally and vertically. Some of this information is obtained by measurement, some by analogy or extrapolation and the rest by guess.

Often the actual waterflood performs even in its early stages in a way quite different from that predicted. The water injectivities do not agree with those predicted, an oil production response is obtained either earlier or later than predicted, and initial water breakthrough occurs perhaps at different wells from those expected. Sometimes differences in predicted and actual performance can be traced to operating problems: casing leaks, plugged perforations, wellbore plugging by solids or bacteria. More often, however, the difference is due to an inadequate reservoir description. Injection surveys can be run to make certain that the injected water is confined to the desired zones. Fluid level surveys will show whether the producers are pumped off.

At this point the wise reservoir engineer will turn again to his prediction method. He will carefully examine the data he used to make his original waterflood performance predictions, and concentrate on those reservoir characteristics that could be of doubtful validity. By prudently adjusting these reservoir characteristics, he can come closer and closer to matching actual injection and production performance. When the important facets of the actual waterflood behavior are matched, the experienced reservoir engineer will be much more confident in the accuracy of his future performance predictions.

This feedback of information from the actual waterflood is an important part of the practical use of waterflood prediction methods. It is precisely this incorporation of actual performance into the prediction technique that makes it possible to forecast with increasing confidence the effects of future changes in injection well location, distribution of injected water between injectors, and water and oil rates.

8.11 Factors Affecting Waterflood Oil Recovery Performance

Any waterflood performance prediction method requires a description of the reservoir. An engineer should attempt to obtain answers to these questions:

1. Is the reservoir likely to perform as a series of independent layers, or as zones of differing permeability with fluid crossflow?
2. Are there zones of high gas saturation or high water saturation that could serve as channels for bypassing water?
3. Does the reservoir contain long natural fractures or directional permeability that could cause preferential areal movement in some direction?
4. Are there areas of high and of low permeability that might cause unbalanced flood performance?
5. Is crossbedding present to the degree that fluid communication between injection and producing wells might be impaired?
6. Is the reservoir likely to contain planes of weakness or closed natural fractures that would open at bottom-hole injection pressures?

Each of these questions represents factors that could cause the reservoir performance to be drastically different from that predicted. The answers to these questions can in many cases be determined by geological and petrophysical studies, which are important and which should be considered as prerequisite to waterflooding operations.

Even in the absence of these factors that cause unfavorable performance, waterfloods frequently recover significantly less oil than predicted. Calloway[65] postulated that the less-than-expected oil recovery was due to resaturation effects. This concept is that the gas space developed in the nonswept portions of the reservoir during primary depletion becomes resaturated with oil during flooding. That is, as the oil bank in the more permeable zones passes by low permeability lenses containing free gas saturation, these lenses become resaturated with oil, and the oil bank is thus reduced or dissipated. These nonswept portions can be lenses or zones of low permeability or that part of the reservoir beyond the last row of producers. There is little doubt that this resaturation occurs in some waterfloods. The two conditions that are necessary for it to take place to the extent that the oil recovery is significantly reduced are (1) that a gas saturation be present prior to waterflooding, and (2) that the reservoir be so heterogeneous that the volumetric sweep is low at the time that the WOR reaches its economic limit. If the reservoir is composed of zones insulated from one another by dense sections or shale, the oil bank formed in one zone cannot resaturate a lower permeability zone by crossflow. Resaturation of tighter zones can occur, however, through the wellbore. To prevent this, producing wells should be kept pumped off.

When maximum oil has been recovered from a reservoir by waterflooding, the residual oil represents the least possible stock-tank oil. This condition exists at the original bubble-point pressure where a barrel of reservoir oil represents the least volume of stock-tank oil. Thus from the standpoint of maximum stock-tank recovery, the optimum pressure at which to flood a reservoir is the original bubble-point pressure. At this pressure also, the reservoir oil viscosity is at its minimum value, which improves the mobility ratio and the areal sweep, and hence the productivity. Other factors favoring waterflooding at the original bubble-point pressure are that (1) the producers have the maximum productivity index; and (2) there is no delay in flood response because, of course, the reservoir is liquid-filled at the start of the flood.

Disadvantages of flooding at the original bubble-point pressure, as compared with starting after some solution gas drive production, are that higher injection pressures are required to inject at the same rates, and that investment in waterflood equipment is required earlier in the field life. This early investment may delay payout, particularly where proration restricts producing rates to a fraction of capacity.

In Chapter 3, Figs. 3.21 through 3.23 show the benefit of trapped gas saturation in improving the waterflood displacement efficiency. However, as noted in that chapter, the pressure increase normally accompanying waterflooding operations results in the solution of any gas trapped in the oil bank, thus eliminating the usefulness of trapped gas in displacing oil.

All this considered, the optimum time to initiate waterflooding is when the reservoir pressure reaches the original saturation pressure.

It is widely accepted that water-wet reservoirs flood better than oil-wet reservoirs. Some have even extended this to say "never waterflood an oil-wet reservoir". This admonishment of course is ridiculous. As previously discussed, the oil recovery from a waterflood depends upon both the displacement efficiency and the volumetric sweep efficiency. Frequently the oil displacement efficiency in an oil-wet reservoir rock is less than that of a water-wet rock having similar pore geometry. It is also true that for the same oil and water viscosities the mobility ratio and hence the volumetric sweep efficiency is less favorable in oil-wet rock than in water-wet rock. Even so, many oil-wet reservoirs have been flooded both efficiently and profitably.

References

1. Schoeppel, R. J.: "Waterflood Prediction Methods", *Oil and Gas J.* (1968) **66**, Jan. 22, 72-75; Feb. 19, 98-106; March 18, 91-93; April 8, 80-86; May 6, 111-114; June 17, 100-105; July 8, 71-79.
2. Yuster, S. T. and Calhoun, J. C., Jr.: "Behavior of Water Injection Wells", *Oil Weekly* (Dec. 18 and 25, 1944) 44-47.
3. Suder, F. E. and Calhoun, J. C., Jr.: "Waterflood Calculations", *Drill. and Prod. Prac.*, API (1949) 260-270.
4. Muskat, M.: "The Effect of Permeability Stratifications in Complete Water Drive Systems", *Trans.*, AIME (1950) **189**, 349-358.
5. Prats, M., Matthews, C. S., Jewett, R. L. and Baker, J. D.: "Prediction of Injection Rate and Production History for Multifluid Five-Spot Floods", *Trans.*, AIME (1959) **216**, 98-105.
6. Fitch, R. A. and Griffith, J. D.: "Experimental and Calculated Performance of Miscible Floods in Stratified Reservoirs". *J. Pet. Tech.* (Nov., 1964) 1289-1298.
7. Stiles, W. E.: "Use of Permeability Distribution in Water Flood Calculations", *Trans.*, AIME (1949) **186**, 9-13.
8. Schmalz, J. P. and Rahme, H. D.: "The Variation of Waterflood Performance with Variation in Permeability Profile", *Prod. Monthly* (Sept., 1950) **15**, No. 9, 9-12.
9. Arps, J. J.: "Estimation of Primary Oil Reserves", *Trans.*, AIME (1956) **207**, 182-191.
10. Ache, P. S.: "Inclusion of Radial Flow in Use of Permeability Distribution in Waterflood Calculations", paper 935-G presented at SPE 32nd Annual Fall Meeting, Dallas, Tex., Oct. 6-9, 1957.
11. Slider, H. C.: "New Method Simplifies Predicting Waterflood Performance", *Pet. Eng.* (Feb., 1961) **33**, B68-78.
12. Johnson, J. P.: "Predicting Waterflood Performance by the Graphical Representation of Porosity and Permeability Distribution", *J. Pet. Tech.* (Nov., 1965) 1285-1290.
13. Dykstra, H. and Parsons, H. L.: "The Prediction of Oil Recovery by Waterflooding", *Secondary Recovery of Oil in the United States*, 2nd ed., API, New York (1950) 160-174.
14. Johnson, C. E., Jr.: "Prediction of Oil Recovery by Water Flood—A Simplified Graphical Treatment of the Dykstra-Parsons Method", *Trans.*, AIME (1956) **207**, 345-346.
15. Felsenthal, M., Cobb, T. R. and Heuer, G. J.: "A Comparison of Waterflood Evaluation Methods", paper SPE 332 presented at SPE Fifth Biennial Secondary Recovery Symposium, Wichita Falls, Tex., May 7-8, 1962.
16. Muskat, M.: *Flow of Homogeneous Fluids Through Porous Systems*, J. W. Edwards, Inc., Ann Arbor, Mich. (1946).
17. Hurst, W.: "Determination of Performance Curves in Five-Spot Waterflood", *Pet. Eng.* (1953) **25**, B40-46.
18. Slobod, R. L. and Caudle, B. H.: "X-Ray Shadowgraph Studies of Areal Sweepout Efficiencies", *Trans.*, AIME (1952) **195**, 265-270.
19. Dyes, A. B., Caudle, B. H. and Erickson, R. A.: "Oil Production After Breakthrough As Influenced by Mobility Ratio", *Trans.*, AIME (1954) **201**, 81-86.
20. Caudle, B. H. and Witte, M. D.: "Production Potential Changes During Sweepout in a Five-Spot Pattern", *Trans.*, AIME (1959) **216**, 446-448.
21. Caudle, B. H. and Loncaric, I. G.: "Oil Recovery in Five-Spot Pilot Floods", *Trans.*, AIME (1960) **219**, 132-136.
22. Kimbler, O. K., Caudle, B. H. and Cooper, H. E., Jr.: "Areal Sweepout Behavior in a Nine-Spot Injection Pattern", *J. Pet. Tech.* (Feb., 1964) 199-202.
23. Caudle, B. H., Hickman, B. M. and Silberberg, I. H.: "Performance of the Skewed Four-Spot Injection Pattern", *J. Pet. Tech.* (Nov., 1968) 1315-1319.
24. Guckert, L. G.: "Areal Sweepout Performance of Seven- and Nine-Spot Flood Patterns", MS thesis, The Pennsylvania State U., University Park (Jan., 1961).
25. Doepel, G. W. and Sibley, W. P.: "Miscible Displacement—A Multilayer Technique for Predicting Reservoir Performance", *J. Pet. Tech.* (Jan., 1962) 73-80.
26. Agan, J. B. and Fernandes, R. J.: "Performance Prediction of a Miscible Slug Process in a Highly Stratified Reservoir", *J. Pet. Tech.* (Jan., 1962) 81-86.
27. Aronofsky, J.: "Mobility Ratio—Its Influence on Flood Patterns During Water Encroachment", *Trans.*, AIME (1952) **195**, 15-24.
28. Aronofsky, J. and Ramey, H. J., Jr.: "Mobility Ratio—Its Influence on Injection and Production Histories in Five-Spot Flood", *Trans.*, AIME (1956) **207**, 205-210.
29. Deppe, J. C.: "Injection Rates—The Effect of Mobility Ratio, Area Swept, and Pattern", *Soc. Pet. Eng. J.* (June, 1961) 81-91.
30. Hauber, W. C.: "Prediction of Waterflood Performance for Arbitrary Well Patterns and Mobility Ratios," *J. Pet. Tech.* (Jan., 1964) 95-103.

31. Buckley, S. E. and Leverett, M. C.: "Mechanism of Fluid Displacement in Sands", *Trans.*, AIME (1942) **146,** 107-116.

32. Felsenthal, M. and Yuster, S. T.: "A Study of the Effect of Viscosity in Oil Recovery by Waterflooding", paper 163-G presented at SPE West Coast Meeting, Los Angeles, Oct. 25-26, 1951.

33. Welge, H. J.: "A Simplified Method for Computing Oil Recovery by Gas or Water Drive", *Trans.*, AIME (1952) **195,** 91-98.

34. Roberts, T. G.: "A Permeability Block Method of Calculating a Water Drive Recovery Factor", *Pet. Eng.* (1959) **31,** B45-48.

35. Kufus, H. B. and Lynch, E. J.: "Linear Frontal Displacement in Multilayer Sands", *Prod. Monthly* (Dec., 1959) **24,** No. 12, 32-35.

36. Snyder, R. W. and Ramey, H. J., Jr.: "Application of Buckley-Leverett Displacement Theory to Non-communicating Layered Systems", *J. Pet. Tech.* (Nov., 1967) 1500-1506.

37. Craig, F. F., Jr., Geffen, T. M. and Morse, R. A.: "Oil Recovery Performance of Pattern Gas or Water Injection Operations from Model Tests", *Trans.*, AIME (1955) **204,** 7-15.

38. Hendrickson, G. E.: "History of the Welch Field San Andres Pilot Waterflood", *J. Pet. Tech.* (Aug., 1961) 745-749.

39. Wasson, J. A. and Schrider, L. A.: "Combination Method for Predicting Waterflood Performance for Five-Spot Patterns in Stratified Reservoirs", *J. Pet. Tech.* (Oct., 1968) 1195-1202.

40. Craig, F. F., Jr.: Discussion of "Combination Method for Predicting Waterflood Performance for Five-Spot Patterns in Stratified Reservoirs", *J. Pet. Tech.* (Feb., 1969) 233-234.

41. Rapoport, L. A., Carpenter, C. W. and Leas, W. J.: "Laboratory Studies of Five-Spot Waterflood Performance", *Trans.*, AIME (1958) **213,** 113-120.

42. Higgins, R. V. and Leighton, A. J.: "A Computer Method to Calculate Two-Phase Flow in Any Irregularly Bounded Porous Medium", *J. Pet. Tech.* (June, 1962) 679-683.

43. Higgins, R. V. and Leighton, A. J.: "Computer Prediction of Water Drive of Oil and Gas Mixtures Through Irregularly Bounded Porous Media—Three-Phase Flow", *J. Pet. Tech.* (Sept., 1962) 1048-1054.

44. Higgins, R. V. and Leighton, A. J.: "Waterflood Prediction of Partially Depleted Reservoirs", paper SPE 757 presented at SPE 33rd Annual California Regional Fall Meeting, Santa Barbara, Oct. 24-25, 1963.

45. Higgins, R. V., Boley, D. W. and Leighton, A. J.: "Aids in Forecasting the Performance of Water Floods", *J. Pet. Tech.* (Sept., 1964) 1076-1082.

46. Higgins, R. V. and Leighton, A. J.: "Computer Techniques for Predicting Three-Phase Flow in Five-Spot Waterfloods", RI 7011, USBM (Aug., 1967).

47. Henley, D. H.: "Method for Studying Waterflooding Using Analog, Digital, and Rock Properties", paper presented at 24th Technical Conference on Petroleum, The Pennsylvania State U., University Park (Oct., 1963).

48. Gurses, B. and Helander, D. P.: "Shape Factor Analysis for Peripheral Waterflood Prediction by the Channel Flow Technique", *Prod. Monthly* (April, 1967) **32,** No. 4, 2-31.

49. Kantar, K. and Helander, D. P.: "Simplified Shape Factor Analysis as Applied to an Unconfined Peripheral Waterflooding System", *Prod. Monthly* (May, 1967) **32,** No. 5, 14-17.

50. Kantar, K. and Helander, D. P.: "Graphical Technique for Interpolating Shape Factors for Peripheral Waterflood Systems", *Prod. Monthly* (June, 1967) **32,** No. 6, 2-5.

51. Douglas, J., Jr., Blair, P. M. and Wagner, R. J.: "Calculation of Linear Waterflood Behavior Including the Effects of Capillary Pressure", *Trans.*, AIME (1958) **213,** 96-102.

52. Hiatt, W. N.: "Injected-Fluid Coverage of Multi-Well Reservoirs With Permeability Stratification", *Drill. and Prod. Prac.*, API (1958) 165-194.

53. Douglas, J., Jr., Peaceman, D. W. and Rachford, H. H., Jr.: "A Method for Calculating Multi-Dimensional Immiscible Displacement", *Trans.*, AIME (1959) **216,** 297-306.

54. Warren, J. E. and Cosgrove, J. J.: "Prediction of Waterflood Behavior in a Stratified System", *Soc. Pet. Eng. J.* (June, 1964) 149-157.

55. Morel-Seytoux, H. J.: "Analytical-Numerical Method in Waterflooding Predictions", *Soc. Pet. Eng. J.* (Sept., 1965) 247-258.

56. Morel-Seytoux, H. J.: "Unit Mobility Ratio Displacement Calculations for Pattern Floods in Homogeneous Medium", *Soc. Pet. Eng. J.*, (Sept., 1966) 217-227.

57. Guthrie, R. K. and Greenberger, M. H.: "The Use of Multiple-Correlation Analyses for Interpreting Petroleum Engineering Data", *Drill. and Prod. Prac.*, API (1955) 130-137.

58. Schauer, P. E.: "Application of Empirical Data in Forecasting Waterflood Behavior", paper 934-G presented at SPE 32nd Annual Fall Meeting, Dallas, Tex., Oct. 6-9, 1957.

59. Guerrero, E. T. and Earlougher, R. C.: "Analysis and Comparison of Five Methods Used to Predict Waterflooding Reserves and Performance", *Drill. and Prod. Prac.*, API (1961) 78-95.

60. Arps., J. J., Brons, F., van Everdingen, A. F., Buchwald, R. W. and Smith, A. E.: "A Statistical Study of Recovery Efficiency" *Bull.* 14D, API (1967).

61. Bush, J. L. and Helander, D. P.: "Empirical Prediction of Recovery Rate in Waterflooding Depleted Sands", *J. Pet. Tech.* (Sept., 1968) 933-943.

62. Wayhan, D. A., Albrecht, R. A., Andrea, D. W. and Lancaster, W. R.: "Estimating Waterflood Recovery in Sandstone Reservoirs", paper 875-24-A presented at Rocky Mountain District Spring Meeting, API Div. of Production, Denver, Colo., April 27-29, 1970.

63. Abernathy, B. F.: "Waterflood Prediction Methods Compared to Pilot Performance in Carbonate Reservoirs", *J. Pet. Tech.* (March, 1964) 276-282.

64. Muskat, M.: *Physical Principles of Oil Production,* McGraw-Hill Book Co., Inc., New York (1949) 682-686.

65. Calloway, F. H.: "Evaluation of Waterflood Prospects", *J. Pet. Tech.* (Oct., 1959) 11-15.

Chapter 9

Pilot Waterflooding

9.1 Advantages and Limitations of Pilot Floods

As discussed in the previous chapters, the prediction of field waterflood performance involves many factors. It is necessary to have reliable information on the displacement efficiency, the areal coverage, and the vertical sweep efficiency. Each of these requires careful sampling to get representative reservoir rock and fluid properties, measures of reservoir heterogeneity, and the like. Thus it is not surprising that enterprising petroleum engineers conceived pilot floods as a means of studying oil recovery performance on an in-place sample of the reservoir itself. This recovery performance could then be scaled up to yield the performance to be expected from full-scale water injection operations.

Economically, a pilot is a desirable tool for estimating field performance. However, it has the following limitations:

1. With a small pilot the probability of locating it in a nonrepresentative portion of the reservoir is increased.
2. The effects of a damaged well will be more pronounced with a small number of wells.
3. Oil migration losses from a single pilot pattern may yield an estimated recovery lower than that to be realized from a full-scale flood.
4. Injected water may be lost outside the pilot area, suggesting a higher injected water requirement than for a full-scale waterflood.

The importance of locating the pilot in a representative portion of the reservoir is obvious. Net pay thickness and oil saturation are the two most important variables that distinguish the oil recovery in one area from that in another. Net pay information is frequently available from cores or logs. Oil saturation, however, can vary even within areas of uniform net pay thickness. In fields having a long history of primary depletion, oil can accumulate in structurally low areas by gravity drainage.

A damaged injection well, or one located in a tight portion of the reservoir can often result in lower-than-anticipated injection rates; and if more than one injector is included in the pilot the result can be an unbalanced flood. A damaged producer has an even more serious effect, with lower recoveries and increased oil migration outside the pilot being the probable results.

Many of the limitations can be overcome by designing pilots to include more than one pattern.

The magnitude of oil migration losses from a pilot will be discussed later.

9.2 Information Obtainable from Pilot Waterfloods

A review of the earliest pilot waterfloods shows that the fundamental objective of these pilots was simply to determine whether or not an oil bank or zone of increased oil saturation could be formed. Consequently, as soon as a "kick" or "buzz" in oil production was obtained, full-scale waterflood development took place. At the same time, of course, water injectivity information was obtained, which was also of value in designing the full-scale flood.

Two general types of pilot waterfloods are run: single five-spot floods and single injection well floods (Fig. 9.1). Some pilot floods comprise multiple patterns— that is, two or more contiguous five-spots—generally to minimize oil migration losses, particularly from an inner pattern.[1]

A number of reservoir engineering studies have been made to permit better evaluation of pilot floods. One of the earliest was that of Paulsell,[2] who found that in a single injection well pilot flood the areal coverage increased after water breakthrough until the swept area constituted 200 percent or more of the pilot area. Thus he concluded that evaluation of a pilot flood based upon the presumption of no flow outside the pilot area would be erroneous.

Rosenbaum and Matthews[3] studied the effect of initial gas saturation and mobility ratio on the ratio of production rate to injection rate for various multiple five-spot patterns. In a study[4] involving both potentiometric and flow models, four different pilot patterns were studied. These included a single five-spot, a single injection well pilot, a cluster of four single injection well pilots, and six inverted five-spots. The ratio of the well diameter to the distance between injection and producing wells was held constant at 1:1,000. A "π ratio", used as a correlating device, was defined as the ratio of the pressure drawdown at the producing wells to the pressure increase at the injection wells. The results of this study showed that as the values of the π ratio increase, the total oil recovery and the total fluid production from the pilot increase relative to the injected water volume. At increased values of the π ratio, oil migrates to the pilot producers from the surrounding area.

Another study of the effect of producing rates and mobility ratio on the performance of a single five-spot pilot flood in a liquid-saturated reservoir was that reported by Caudle and Loncaric.[5] Ratios of injection to producing rates were varied from 1 to 4 and mobility ratios ranged from 0.1 to 10. At low rate ratios, the pilot producers recovered up to four times the recoverable oil originally within the pilot area. Thus under these rate conditions oil migrated into the pilot area and an overly optimistic estimate of the oil recovery possible by full-scale flooding was obtained.

The effect of an initial gas saturation on single injection well pilot flooding was studied by Neilson and Flock.[6] After water breakthrough, the wells continued to produce oil until ultimate areal sweep efficiencies up to 600 percent were obtained.

Movement of oil *into* the pilot producers is rarely experienced in the field. More frequently the oil recovery from a pilot is lower than that obtained by full-scale flooding because of migration *out of* the pilot. The possible magnitude of oil migration from the pilot was discussed in a paper[7] that reported the results of a study of well productivity on the performance of both single five-spot and single injection well pilot floods. The productivity of the producing well was expressed by its Condition Ratio. This term is defined as the ratio of the flow capacity of the reservoir as determined from productivity index data to the flow capacity determined from pressure buildup data. This is equivalent to the ratio of the actual well productivity to that of an undamaged, nonstimulated, normal-sized well in the same formation. A normal-sized well is taken to be a 6-in.-diameter well on 10-acre spacing. Mobility ratios of 0.20, 0.45, and 0.94 were studied with gas saturations up to 25.9 percent PV. The results of this study are discussed in the following section.

Single Five-Spot Pilot

During fillup of the gas space originally in the pilot area, there was radial movement of the injected water and no oil migration across the pilot area boundaries. After liquid fillup, if the Condition Ratio of the pilot producer was not sufficient to produce all the pilot area oil displaced, the oil would migrate out of the pilot area. One factor tending to retard the migration of oil from the pilot area is the presence of waterflood fronts formed about each of the injectors. As these water fronts come closer to coalescing, less recoverable oil within the pilot area can escape. The model tests indicated that as gas saturations increased, a greater portion of the recoverable oil was produced from the pilot producer. At higher gas saturations, the pilot area was more completely enclosed by the waterflood fronts at the time of liquid fillup within the pilot area.

The initial gas saturation could be so great that, at the time of liquid fillup within the pilot area, the waterflood fronts would coalesce, and thus oil migration would be prevented. The value of this initial gas saturation is

$$S_g^* = \frac{\pi}{4}(1 - S_{wc} - S_{or}) \quad \ldots \ldots \quad (9.1)$$

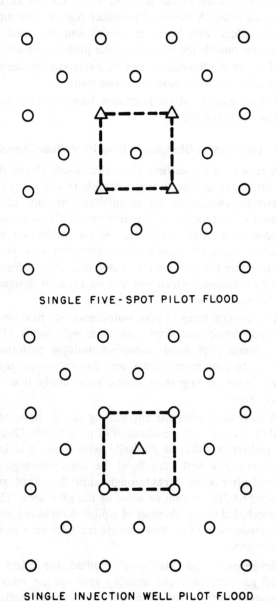

Fig. 9.1 Two frequently used pilot waterflood patterns.

Model results indicating how initial gas saturation and producing well Condition Ratio affect the oil recovery from a single five-spot pilot flood are shown in Fig. 9.2. At a Condition Ratio of 2.22, equivalent to a high-capacity horizontal fracture with a radius of 18.5 ft, approximately 93 percent of the recoverable oil originally in the pilot area was produced. The range of values of $S_{gi}/S_g{}^*$ encountered in conventional waterflooding operations is from 0.3 to 0.7. In this range of gas saturations, a single five-spot pilot flood with a producing well Condition Ratio of 1.0 (equivalent to a clean, normal-sized well having no production stimulation) will recover from 57 to 73 percent of the recoverable oil originally in place. Thus for a pilot to yield oil recovery representative of a full-scale pattern waterflood, the producing wells should have Condition Ratios of 2.22 or more.

Fig. 9.3 presents the oil recovery performance obtained from a developed five-spot and from a single five-spot pilot flood in which the pilot producer has a Condition Ratio of 2.22. The agreement confirms that at a producing Condition Ratio of 2.22, a single five-spot pilot flood can also yield a good estimate of oil production performance expected under full-scale flooding. Regional pressure gradients were also shown to have little effect on pilot performance.[7]

Single Injection Well Pilot Flooding

Ref. 7 discusses the results of single injection well pilot model tests. The study showed that when there was no gas saturation existing before water injection, the oil recovered by the pilot producers approximated the volume of recoverable oil in the pilot area when the producers had a Condition Ratio of 1.0 or above. However, when an initial gas saturation was present, there was significant oil migration outside the pilot area regardless of the Condition Ratio of the producing wells. A single injection well pilot flood is generally inadequate for obtaining an estimate of full-scale waterflood recovery. However, from the injected water volume at water breakthrough, useful information can be obtained on volumetric coverage at breakthrough. Also, if water breaks through at one offset producer earlier than at the other offsets, directional permeability will be reflected.

Table 5.6 and Fig. 5.4 show the areal sweep efficiency of a single injection well pilot flood at water breakthrough as it depends upon mobility ratio. Complete or nearly complete areal coverage of the pilot is obtained at water breakthrough.

9.3 Pilot Flood Design

To properly design a pilot flood, one must bear in mind the objectives of such a pilot. If the objectives are to determine the water injectivity and to obtain an indication of a substantial movable oil volume, almost any pilot flood pattern is suitable. However, if the objectives also include obtaining an estimate of the recoverable oil by waterflooding, then the pilot must

1. Be located in a portion of the reservoir representative of the oil saturation, permeability, and reservoir heterogeneity of the remainder of the reservoir,
2. Be composed of a single or multiple five-spot pattern, with the producers stimulated so that they have a Condition Ratio of 2.22 or above, and

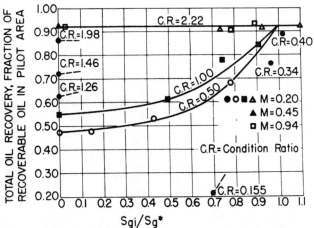

Fig. 9.2 Effect of producing well Condition Ratio and initial gas saturation on oil recovery of a single five-spot pilot waterflood — four injectors, and one producer.[7]

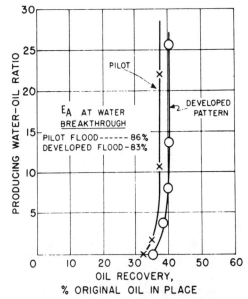

Fig. 9.3 Model pilot and developed pattern WOR performance; no initial gas saturation, single five-spot pilot flood.[7]

3. Have injection rates for each injector proportional to the product of the porosity and net pay thickness of the areas surrounding each injector.

These are not simple requirements, and frequently a great deal of effort is involved in satisfying them.

References

1. Bernard, W. J. and Caudle, B. H.: "Model Studies of Pilot Waterfloods", *J. Pet. Tech.* (March, 1967) 404-410.
2. Paulsell, B. L.: "Areal Sweep Performance of Five-Spot Pilot Floods", MS thesis, The Pennsylvania State U., University Park (Jan., 1958).
3. Rosenbaum, M. J. F. and Matthews, C. S.: "Studies on Pilot Water Flooding", *Trans.*, AIME (1959) **216**, 316-323.
4. Dalton, R. L., Jr., Rapoport, L. A. and Carpenter, C. W., Jr.: "Laboratory Studies of Pilot Water Floods", *Trans.*, AIME (1960) **219**, 24-30.
5. Caudle, B. H. and Loncaric, I. G.: "Oil Recovery in Five-Spot Pilot Floods", *Trans.*, AIME (1960) **219**, 132-136.
6. Neilson, I. D. R. and Flock, D. L.: "The Effect of a Free Gas Saturation on the Sweep Efficiency of an Isolated Five-Spot", *Bull.*, CIM (1962) **55**, 124-129.
7. Craig, F. F., Jr.: "Laboratory Model Study of Single Five-Spot and Single Injection-Well Pilot Waterflooding", *J. Pet. Tech.* (Dec., 1965) 1454-1460.

Chapter 10

Conclusion

10.1 The State of the Art

Waterflooding is a proven oil recovery process. It is not always successful and economically profitable, but there is a sound basis for engineering waterflood projects. Problems associated with the measurement of basic water-oil flow properties have largely been solved. The significance of various types of reservoir nonuniformities on areal and vertical coverage is well known. In addition, methods of predicting waterflood performance have reached a high degree of sophistication. Thus we have at our disposal techniques for good engineering appraisal of waterflooding.

10.2 Current Problems and Areas for Further Study

Description of Reservoir Heterogeneities

With our present and growing ability to calculate accurately the performance of a reservoir, given a detailed description of its heterogeneities, the need for improved description of these nonuniformities for specific fields has become more acute. The goal of good waterflood engineering can be reached only by a symbolic step-ladder, one leg of which is composed of the waterflood prediction methods and the other, of a *quantitative* description of reservoir nonuniformities.

This need for quick and accurate description of reservoir heterogeneities becomes more critical as the industry turns its attention to less permeable, naturally fractured reservoirs as waterflood candidates. Here the key to sound appraisal is a knowledge of the location and magnitude of these fractures. In reservoirs of low permeability, it is generally necessary to fracture water injection wells to obtain injection rates high enough to support economic oil producing rates. Unfortunately these high injection rates frequently open "planes of weakness" or joint systems that provide ready paths through which water can bypass and channel. We have much to learn about waterflooding low-permeability reservoirs.

Improved Waterflooding Methods

With the growing understanding of waterflooding, more and more attention has been devoted to the development of improved waterflooding methods.

To assist oil recovery in reservoirs, such as the Spraberry Trend, in which reservoir scale fractures exist, pressure-pulsing with water has been tried.[1] This involves periods of water injection and pressuring, followed by periods of depressuring, with little or no water injection. Laboratory studies[2] have shown that this process, applicable to both preferentially water-wet and preferentially oil-wet rocks, can yield additional recoveries in the range of 5 to 10 percent pore space.

The oil recovery performance of waterfloods can be improved by methods that increase the oil displacement efficiency, the volumetric sweep efficiency, or both. These improved methods, which involve the addition of gas, chemicals, solvent, or heat (Fig. 10.1) to conventional waterflooding, will be touched on briefly.

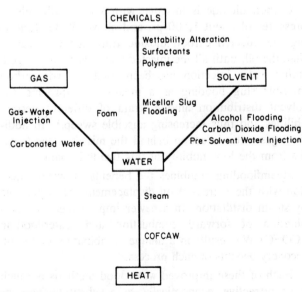

Fig. 10.1 Improved waterfloods.

The injection of gas along with water decreases the water mobility, improving volumetric coverage. An additional benefit is the increased oil displacement efficiency due to trapped gas.[3] When carbon dioxide is injected with water, the resulting reduced oil viscosity and crude oil swelling can yield increased oil displacement.[4-6] Foam injection involves the injection of air, water, and a chemical agent that stabilizes the foam.[7] Foam is essentially a very viscous, aqueous-appearing fluid. Oil recovery benefits resulting from its use are those of a thick water. Foam flooding is still in the development stage.

Another means for increasing the effective viscosity of water is the addition of a polymer. Polymer solutions have shown promise in both laboratory and field testing.[8] In one of the first attempts to make water miscible with oil by decreasing the water-oil interfacial tension, detergents[9] were added to the injected water. The loss by adsorption on the reservoir rock surfaces has severely limited the usefulness of detergents. Wettability alteration[10] involves the injection of a chemical designed to alter the rock wettability from preferentially oil-wet to preferentially water-wet. Field testing[11] has shown that wettability alteration can increase displacement efficiency and move oil that has been left behind after waterflooding.

In a recent development, micellar solutions[12] composed of micro-emulsions of oil, water, alcohol, and surfactants show great promise as a miscible slug material to separate oil from the injected water, resulting in miscible displacement. Studies have been conducted[13] in which alcohol was injected as a solvent slug between the oil and the following water in an attempt to cause miscible displacement. Use of isopropyl alcohol appears limited because it is initially miscible with both the oil and connate water, so that the alcohol content in the mixing zone is diluted below that level necessary to maintain miscibility. The use of multiple alcohol slugs appears promising but is more costly.

Carbon dioxide is miscible with some crude oils at pressures of about 1,500 psi and above. Water following the injected carbon dioxide traps the CO_2 rather than the oil, until all the injected CO_2 has been trapped. Initial water injection has been used in hydrocarbon miscible slug flooding as a means of improving the solvent distribution among strata of different permeabilities and thus increasing miscible sweep.[14] In addition, gas-water injection behind the miscible slug benefits from the low mobility of this two-fluid bank.[15]

Steamflooding combines the benefits of water injection with the increased oil displacement made possible by steam distillation. In a newer improvement, a combination of forward combustion and waterflooding (COFCAW) results in a unique combination of the oil recovery benefits of each process.[16]

Each of these improved waterflood methods is much more attractive, economically, as a substitute for conventional waterflooding than as a means for tertiary recovery.

References

1. Elkins, L. F. and Skov, A. M.: "Cyclic Waterflooding the Spraberry Utilizes 'End Effects' to Increase Oil Production Rate", *J. Pet. Tech.* (Aug., 1963) 877-883.
2. Owens, W. W. and Archer, D. L.: "Waterflood Pressure Pulsing for Fractured Reservoirs", *J. Pet. Tech.* (June, 1966) 745-752.
3. Pfister, R. J.: "More Oil From Spent Water Drives by Intermittent Air or Gas Injection", *Prod. Monthly* (Sept., 1947) **12**, No. 9, 10-12.
4. Beeson, D. M. and Ortloff, G. D.: "Laboratory Investigation of the Water-Driven Carbon Dioxide Process for Oil Recovery", *Trans.*, AIME (1959) **216**, 388-391.
5. Hickok, C. W., Christensen, R. J. and Ramsay, H. J., Jr.: "Progress Review of the K&S Carbonated Waterflood Project", *J. Pet. Tech.* (Dec., 1960) 20-24.
6. Scott, J. O. and Forrester, C. E.: "Performance of Domes Unit Carbonated Waterflood—First Stage", *J. Pet. Tech.* (Dec., 1965) 1379-1384.
7. Fried, A. N.: "The Foam-Drive Process for Increasing the Recovery of Oil", RI 5866, USBM (1961).
8. Sandiford, B. B.: "Laboratory and Field Studies of Water Floods Using Polymer Solutions to Increase Oil Recoveries", *J. Pet. Tech.* (Aug., 1964) 917-922.
9. Taber, J. J.: "The Injection of Detergent Slugs in Water Floods", *Trans.*, AIME (1958) **213**, 186-192.
10. Leach, R. O. and Wagner, O. R.: "Improving Oil Displacement Efficiency by Wettability Adjustment", *Trans.*, AIME (1959) **216**, 65-72.
11. Leach, R. O., Wagner, O. R., Wood, H. W. and Harpke, C. F.: "A Laboratory and Field Study of Wettability Adjustment in Waterflooding", *J. Pet. Tech.* (Feb., 1962) 205-212.
12. Gogarty, W. B. and Tosch, W. C.: "Miscible-Type Waterflooding: Oil Recovery with Micellar Solutions", *J. Pet. Tech.* (Dec., 1968) 1407-1414.
13. Gatlin, C. and Slobod, R. L.: "The Alcohol Slug Process for Increasing Oil Recovery", *Trans.*, AIME (1960) **219**, 46-53.
14. Fitch, R. A. and Griffith, J. D.: "Experimental and Calculated Performance of Miscible Floods in Stratified Reservoirs", *J. Pet. Tech.* (Nov., 1964) 1289-1298.
15. Caudle, B. H. and Dyes, A. B.: "Improving Miscible Displacement by Gas-Water Injection", *Trans.*, AIME (1958) **213**, 281-284.
16. Parrish, D. R. and Craig, F. F., Jr.: "Laboratory Study of a Combination of Forward Combustion and Waterflooding—The COFCAW Process", *J. Pet. Tech.* (June, 1969) 753-761.

Appendix A

Derivation of Fractional Flow Equation

In deriving the fractional flow equation for the case of water displacing oil, Darcy's law is written separately for oil flow and for water flow, in consistent units, as follows:

$$u_o = -\frac{k_o}{\mu_o}\left(\frac{\partial p_o}{\partial L} + g\,\rho_o \sin \alpha_d\right), \quad \text{(A-1)}$$

$$u_w = -\frac{k_w}{\mu_w}\left(\frac{\partial p_w}{\partial L} + g\,\rho_w \sin \alpha_d\right) \quad \text{(A.2)}$$

Transforming Eqs. A.1 and A.2 yields

$$u_o \frac{\mu_o}{k_o} = -\frac{\partial p_o}{\partial L} - g\,\rho_o \sin \alpha_d, \quad \text{(A.1a)}$$

$$u_w \frac{\mu_w}{k_w} = -\frac{\partial p_w}{\partial L} - g\,\rho_w \sin \alpha_d \quad \text{(A.2a)}$$

Subtracting Eq. A.1a from Eq. A.2a, we obtain

$$u_w \frac{\mu_w}{k_w} - u_o \frac{\mu_o}{k_o} = -\left(\frac{\partial p_w}{\partial L} - \frac{\partial p_o}{\partial L}\right) - g(\rho_w - \rho_o)\sin \alpha_d \quad \text{(A.3)}$$

The capillary pressure is defined as the pressure in the oil phase minus the pressure in the water phase. Hence,

$$P_c = p_o - p_w \quad \text{(A.4)}$$

The density difference is here defined as the difference between the water density and the oil density. Thus,

$$\Delta \rho = \rho_w - \rho_o \quad \text{(A.5)}$$

Substituting Eqs. A.4 and A.5 into Eq. A.3, we obtain

$$u_w \frac{\mu_w}{k_w} - u_o \frac{\mu_o}{k_o} = \frac{\partial P_c}{\partial L} - g\,\Delta\rho \sin \alpha_d. \quad \text{(A.6)}$$

The total velocity, u_t, can be defined as the sum of the water and oil velocities. Thus,

$$u_t = u_o + u_w \quad \text{(A.7)}$$

Substituting Eq. A.7 into Eq. A.6 to eliminate the oil velocity, u_o, we obtain,

$$u_w \frac{\mu_w}{k_w} - (u_t - u_w)\frac{\mu_o}{k_o} = \frac{\partial P_c}{\partial L} - g\,\Delta\rho \sin \alpha_d, \quad \text{(A.8)}$$

or

$$u_w\left(\frac{\mu_w}{k_w} + \frac{\mu_o}{k_o}\right) - u_t \frac{\mu_o}{k_o} = \frac{\partial P_c}{\partial L} - g\,\Delta\rho \sin \alpha_d. \quad \text{(A.8a)}$$

Dividing Eq. A.8a by the total velocity, u_t, gives

$$\frac{u_w}{u_t}\left(\frac{\mu_w}{k_w} + \frac{\mu_o}{k_o}\right) - \frac{\mu_o}{k_o} = \frac{1}{u_t}\left(\frac{\partial P_c}{\partial L} - g\,\Delta\rho \sin \alpha_d\right) \quad \text{(A.9)}$$

Solving Eq. A.9 for the term u_w/u_t yields

$$\frac{u_w}{u_t} = \frac{\dfrac{\mu_o}{k_o} + \dfrac{1}{u_t}\left(\dfrac{\partial P_c}{\partial L} - g\,\Delta\rho \sin \alpha_d\right)}{\dfrac{\mu_w}{k_w} + \dfrac{\mu_o}{k_o}} \quad \text{(A.9a)}$$

The term f_w is defined as the fraction of water in the total flowing stream. Thus by definition,

$$f_w = \frac{u_w}{u_t}. \quad \text{(A.10)}$$

Substituting Eq. A.10 in Eq. A.9a, and dividing both numerator and denominator of the right side of Eq. A.9a by the term μ_o/k_o we obtain

$$f_w = \frac{1 + \dfrac{k_o}{u_t \mu_o}\left(\dfrac{\partial P_c}{\partial L} - g\,\Delta\rho \sin \alpha_d\right)}{1 + \dfrac{\mu_w}{\mu_o}\dfrac{k_o}{k_w}}. \quad \text{(A.11)}$$

Eq. A.11, which is the complete form of the fractional flow equation, is also shown as Eq. 3.1. It expresses the fraction of the total water in the total flowing stream in terms of the fluid viscosities, the water and oil relative permeabilities, the total fluid velocity, the capillary pressure gradient, and gravitational forces.

Consider the special case in which the displacement of oil by water occurs in a horizontal system. Also let us assume that the capillary pressure gradient is negligible. Then Eq. A.11 reduces to

$$f_w = \frac{1}{1 + \frac{\mu_w}{\mu_o} \frac{k_o}{k_w}} \quad \ldots \ldots \ldots \quad (A.12)$$

Eq. A.12, known as the simplified form of the fractional flow equation, is also shown as Eq. 3.4.

It is obvious that the same approach used here could be used to derive the fractional flow equation applying to gas displacement of oil. In the resulting form of Eqs. A.11 and A.12, each subscript w, indicating water, would be replaced by a subscript g, signifying gas.

Appendix B

Derivation of the Frontal Advance Equation

The derivation of the frontal advance equation involves only two assumptions: (1) there is no mass transfer between the phases, and (2) the phases are incompressible.

Consider an infinitesimal element of rock having a porosity ϕ, an area A, and a length ΔL, in the direction of flow. The mass rate of water—the displacing fluid entering the element—at point L is

$(q_w \rho_w)_L.$

The mass rate of water leaving the element at point $L + \Delta L$ is

$(q_w \rho_w)_{L + \Delta L}.$

The mass rate of water accumulation in the element is then

$A \phi \Delta L \dfrac{\partial}{\partial t} (S_w \rho_w).$

The mass rate of water entering the element of rock minus the mass rate of water leaving is by material balance principles equal to the rate of mass accumulation of water in the rock element. Thus

$$(q_w \rho_w)_L - (q_w \rho_w)_{L + \Delta L} = A \phi \Delta L \frac{\partial}{\partial t} (S_w \rho_w)$$
$$\quad \ldots \ldots \ldots (B.1)$$

By definition,

$$F(L, t) - F(L + \Delta L, t) = -\left(\frac{\partial F}{\partial L}\right)_t dL . \quad (B.2)$$

So then Eq. B.1 can be written

$$\frac{\partial}{\partial L}(q_w \rho_w) + A \phi \frac{\partial}{\partial t}(S_w \rho_w) = 0 \quad . \quad . \quad (B.3)$$

However, the fluids are considered to be incompressible. Thus the water density, ρ_w, is not a function of either time or distance. So

$$\frac{\partial}{\partial L}(q_w) + A \phi \frac{\partial}{\partial t}(S_w) = 0 \quad , \quad . \quad . \quad (B.3a)$$

and

$$\left(\frac{\partial S_w}{\partial t}\right)_L = - \frac{1}{A\phi} \left(\frac{\partial q_w}{\partial L}\right)_t . \quad . \quad . \quad . \quad (B.3b)$$

Since the water flow rate, q_w, is a function of both the water saturation and time,

$$dq_w = \left(\frac{\partial q_w}{\partial S_w}\right)_t dS_w + \left(\frac{\partial q_w}{\partial t}\right)_{S_w} dt \quad . \quad . \quad (B.4)$$

Taking the derivative with respect to length, L, at a fixed time, t, yields

$$\left(\frac{\partial q_w}{\partial L}\right)_t = \left(\frac{\partial q_w}{\partial S_w}\right)_t \left(\frac{\partial S_w}{\partial L}\right) \quad , \quad . \quad . \quad . \quad (B.5)$$

or

$$\left(\frac{\partial S_w}{\partial L}\right)_t = \left(\frac{\partial q_w}{\partial L}\right)_t \Big/ \left(\frac{\partial q_w}{\partial S_w}\right)_t . \quad . \quad . \quad (B.5a)$$

Similarly, the water saturation, S_w, is a function of both distance and time, so at a constant saturation

$$dS_w = \left(\frac{\partial S_w}{\partial L}\right)_t dL + \left(\frac{\partial S_w}{\partial t}\right)_L dt = 0 \quad . \quad (B.6)$$

Thus

$$\left(\frac{\partial L}{\partial t}\right)_{S_w} = -\left(\frac{\partial S_w}{\partial t}\right)_L \Big/ \left(\frac{\partial S_w}{\partial L}\right)_t . \quad . \quad (B.6a)$$

Substituting Eqs. B.3b and B.5a into Eq. B.6a, we obtain

$$\left(\frac{\partial L}{\partial t}\right)_{S_w} = \frac{1}{A\phi} \left(\frac{\partial q_w}{\partial S_w}\right)_t . \quad . \quad . \quad . \quad . \quad (B.7)$$

The term f_w is by definition the fraction of water in the total flowing stream. Thus

$$q_w = f_w q_t \quad . \quad . \quad . \quad . \quad . \quad . \quad . \quad . \quad (B.8)$$

Differentiating Eq. B.8 with respect to water saturation, S_w, at constant time, t,

$$\left(\frac{\partial q_w}{\partial S_w}\right)_t = f_w \left(\frac{\partial q_t}{\partial S_w}\right)_t + q_t \left(\frac{\partial f_w}{\partial S_w}\right)_t \quad . \quad (B.9)$$

However, since the fluids are incompressible, the change in total velocity with saturation at any time is zero. Thus

$$\left(\frac{\partial q_w}{\partial S_w}\right)_t = q_t \left(\frac{\partial f_w}{\partial S_w}\right)_t \quad . \quad . \quad . \quad . \quad (B.9a)$$

Substituting Eq. B.9a into Eq. B.7 yields

$$\left(\frac{\partial L}{\partial t}\right)_{S_w} = \frac{q_t}{A\phi} \left(\frac{\partial f_w}{\partial S_w}\right)_t \quad . \quad . \quad . \quad . \quad (B.10)$$

This equation states that the rate of advance, or velocity, of a plane of saturation is equal to the superficial total fluid velocity multiplied by the derivative of the fractional flow with respect to water saturation.

Appendix C

Alternative Derivation of Welge's Equation

In 1952, Welge* presented a derivation of the equation relating the average and producing end saturations during a gas or water drive. In this derivation it is implied that the slope of the fractional flow curve, df/dS, also denoted by the term f', is zero at the maximum injected fluid saturation. As discussed in Chapter 3, the value of f' at the maximum displacing fluid saturation is not zero, but has a finite value. In this Appendix, we shall show that Welge's equation also can be derived for the condition of a nonzero value of f'.

Consider a one-dimensional system having a saturation gradient shown in Fig. C.1. The average displacing fluid saturation, $\overline{S}$, can be evaluated as follows:

$$\overline{S} = \frac{S_1 x_1 + \int_1^2 S\, dx}{x_2} \quad \ldots \ldots \quad (C.1)$$

Since the distance, x, attained in the rock by any given saturation is proportional to f', Eq. C.1 becomes

$$\overline{S} = \frac{S_1 f_1' + \int_1^2 S\, df'}{f_2'} \quad \ldots \ldots \quad (C.2)$$

Upon integrating by parts, Eq. C.2 becomes

$$\overline{S} = \frac{S_1 f_1' + S_2 f_2' - S_1 f_1' - \int_1^2 f'\, dS}{f_2'} ; \quad (C.3)$$

*Welge, H. J.: "A Simplified Method for Computing Oil Recovery by Gas or Water Drive", Trans., AIME (1952) 195, 91-98.

or

$$\overline{S} = \frac{S_2 f_2' - (f_2 - 1)}{f_2'}. \quad \ldots \ldots \quad (C.3a)$$

Taking Eq. B.10, separating variables, and substituting f' for df_w/dS_w, we obtain

$$dL = \frac{q_t}{A\phi} f'\, dt \quad \ldots \ldots \quad (C.4)$$

Integrating Eq. C.4 and expressing the results in terms of the exit end of the system yields

$$L = \frac{f_2'}{A\phi} \int q_t\, dt = \frac{f_2'}{A\phi} W_i ,$$

or

$$f_2' = \frac{LA\phi}{W_i} = \frac{1}{Q_i}, \quad \ldots \ldots \quad (C.5)$$

where Q_i is the pore volumes of cumulative injected fluid. Also

$$f_{o2} = 1 - f_2 \quad \ldots \ldots \quad (C.6)$$

Substituting Eqs. C.5 and C.6 into Eq. C.3a yields

$$\overline{S} = S_2 + Q_i f_{o2} \quad \ldots \ldots \quad (C.7)$$

Eq. C.7 is identical with that developed by Welge and shown as Eq. 3.10 in Chapter 3.

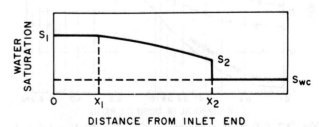

Fig. C.1 Saturation gradient during waterflood.

Appendix D

Additional Published Design Charts and Correlations

Figs. D.1 through D.6 are taken from
Dyes, A. B., Caudle, B. H., and Erickson, R. A.: "Oil Production After Breakthrough — As Influenced by Mobility Ratio", *Trans.*, AIME (1954) **201**, 81-86.

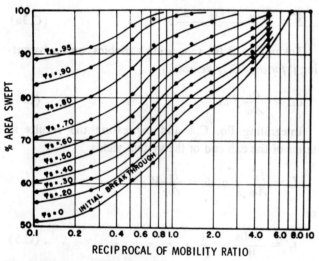

Fig. D.1 Effect of mobility ratio on oil production for the five-spot pattern.

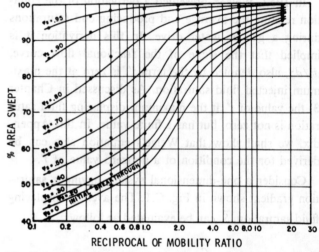

Fig. D.3 Effect of mobility ratio on oil production for the direct line drive (square pattern); $d/a = 1$.

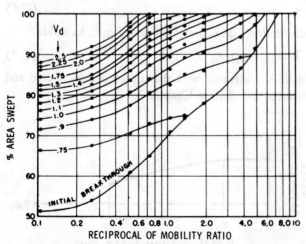

Fig. D.2 Effect of mobility ratio on the displaceable volumes injected for the five-spot pattern.

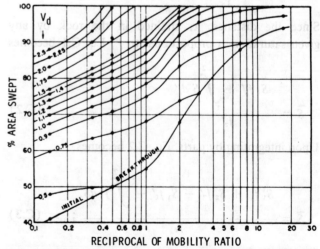

Fig. D.4 Effect of mobility ratio on the displaceable volumes injected for the direct line drive (square pattern); $d/a = 1$.

APPENDIX D

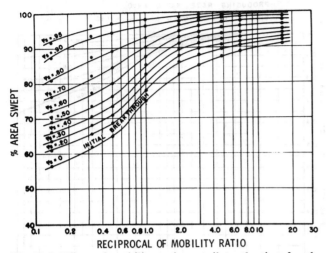

Fig. D.5 Effect of mobility ratio on oil production for the staggered line drive; $d/a=1$.

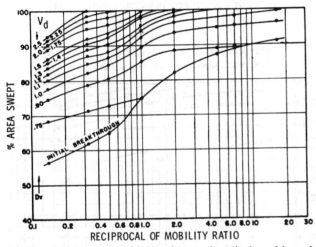

Fig. D.6 Effect of mobility ratio on the displaceable volumes injected for the staggered line drive; $d/a=1$.

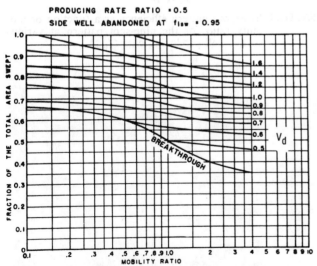

Fig. D.7 Sweepout pattern efficiency as a function of mobility ratio for the nine-spot pattern at various displaceable volumes injected.

Figs. D.7 through D.12 are taken from Kimbler, O. K., Caudle, B. H., and Cooper, H. E., Jr.: "Areal Sweepout Behavior in a Nine-Spot Injection Pattern", *Trans.*, AIME (1964) **231**, 199-202.

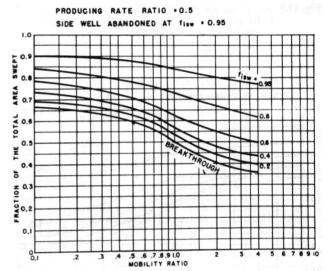

Fig. D.8A Sweepout pattern efficiency as a function of mobility ratio for the nine-spot pattern at various side-well producing cuts (f_{isw}).

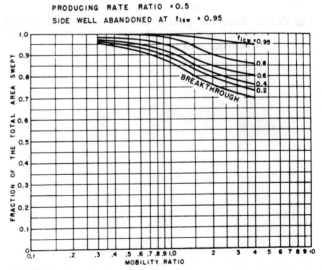

Fig. D.8B Sweepout pattern efficiency as a function of mobility ratio for the nine-spot pattern at various corner-well producing cuts (f_{icw}).

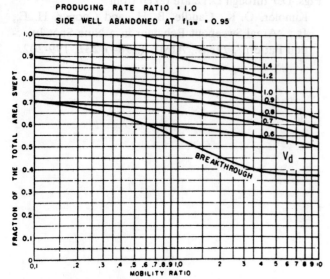

Fig. D.9 Sweepout pattern efficiency as a function of mobility ratio for the nine-spot pattern at various displaceable volumes injected.

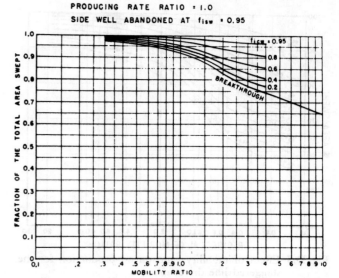

Fig. D.10B Sweepout pattern efficiency as a function of mobility ratio for the nine-spot pattern at various corner-well producing cuts (f_{icw}).

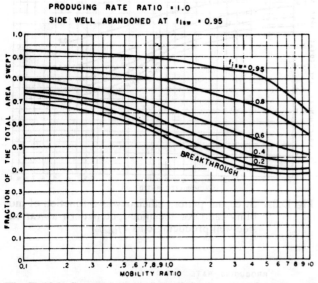

Fig. D.10A Sweepout pattern efficiency as a function of mobility ratio for the nine-spot pattern at various side-well producing cuts (f_{isw}).

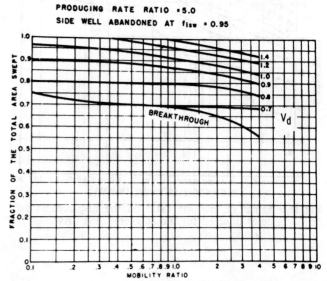

Fig. D.11 Sweepout pattern efficiency as a function of mobility ratio for the nine-spot pattern at various corner-well producing cuts (f_{icw}).

APPENDIX D

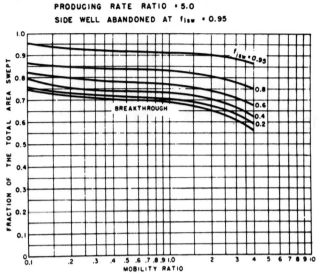

Fig. D.12A Sweepout pattern efficiency as a function of mobility ratio for the nine-spot pattern at various side-well producing cuts (f_{isw}).

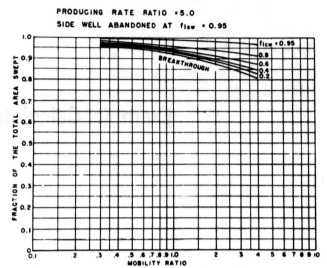

Fig. D.12B Sweepout pattern efficiency as a function of mobility ratio for the nine-spot pattern at various corner-well producing cuts (f_{icw}).

Fig. D.13 is taken from
"How Temperature Affects Viscosity of Salt Water", *World Oil* (Aug. 1, 1967) 68.

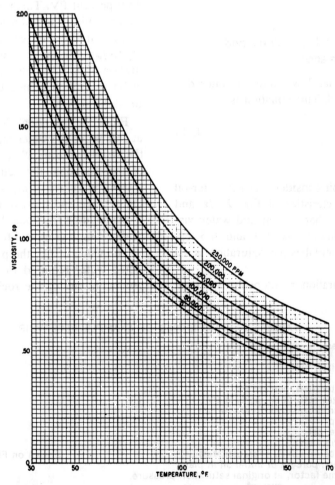

Fig. D.13 How temperature affects viscosity of salt water. Curves indicate the effect of temperature on viscosity of salt water solutions of various concentrations. Chart may be used for estimating viscosity of salt water when calculating mobility ratios.

Appendix E

Example Calculations

The example reservoir has the properties shown on Table E.1. Calculations will be made, with data from an example reservoir, to illustrate the use of various types of waterflood performance prediction methods. This reservoir has been subject to some pressure depletion by solution gas drive, then to gas injection for a brief period of time.

E.1 Calculation of Fractional Flow Curve and Displacement Performance

To calculate the fractional flow curve for water displacing oil, Eq. 3.4a is used. This equation is

$$f_w = \frac{1}{1 + \frac{\mu_w}{k_{rw}} \frac{k_{ro}}{\mu_o}} \qquad \qquad (E.1)$$

For the example reservoir consider that the water-oil relative permeability characteristics of Fig. 2.17a and 2.17b apply. Consider also that the oil and water viscosities at reservoir conditions are 1.0 and 0.5 cp, respectively, and that the initial water saturation is 10 percent PV.

The fractional flow-saturation relationship is shown in Table E.2.

The calculated fractional flow curve, shown in Fig. E.1, is identical with those in Figs. 3.2 and 3.8.

A tangent is drawn to the fractional flow curve from the initial water saturation, 10 percent PV (see Fig. E.1). The water saturation at the tangent is 0.469 or 46.9 percent PV. This is the water saturation at the upstream end of the stabilized zone, S_{wsz}. If we extend the tangent until it intersects the horizontal line corresponding to $f_w = 1.0$, the intersection will occur at a water saturation of 0.563 or 56.3 percent PV. This is the average water saturation behind the flood front at and prior to breakthrough, $\bar{S}_{wbt}$.

From the fractional flow curve for water saturations at and above the water saturation at the upstream end of the stabilized zone, S_{wsz} (46.9 percent PV), the slope of the fractional flow curve is determined. Table E.3 shows the values of df_w/dS_w tabulated along with both the average and the exit-end water saturations and the corresponding fractional flow. Fig. E.2 shows the values of df_w/dS_w in graphical form.

The value of Q_i, the pore volumes of cumulative injected water required to reach a saturation S_{w2} at the producing end of the rock system[1] at and after break-

TABLE E.1 — PROPERTIES OF EXAMPLE RESERVOIR

Well spacing, acres	20
Thickness, ft	50
Average permeability, md	10
Porosity, percent	20
Connate water saturation, percent PV	10
Current average gas saturation, percent PV	15
Oil viscosity at current reservoir pressure, cp	1.0
Water viscosity, cp	0.5
Reservoir pressure, psi	1,000
Permeability distribution	Shown on Fig. E.3
Water-oil relative permeability characteristics	Shown on Figs. 2.17a and 2.17b
Current oil recovery, percent of initial oil in place	10.4
Oil reservoir volume factor, at original saturation pressure	1.29
current	1.20
Flooding pattern	Five-spots using existing wells
Pattern area, acres*	40
Wellbore radius, ft	1.0

*Complete five-spot pattern contains two total wells, one injector and one producer.

APPENDIX E

TABLE E.2 — FRACTIONAL FLOW CURVE, EXAMPLE PROBLEM

Water Saturation, S_w (fraction)	Relative Permeability Oil, k_{ro} (fraction)	Water, k_{rw} (fraction)	Fractional Flow of Water, f_w
0.10	1.000	0.000	0.0000
0.30	0.373	0.070	0.2729
0.40	0.210	0.169	0.6168
0.45	0.148	0.226	0.7533
0.50	0.100	0.300	0.8571
0.55	0.061	0.376	0.9250
0.60	0.033	0.476	0.9665
0.65	0.012	0.600	0.9901
0.70	0.000	0.740	1.0000

TABLE E.3 — WATERFLOOD DISPLACEMENT PERFORMANCE, EXAMPLE PROBLEM

S_{w2}, Exit-End Water Saturation (fraction PV)	f_{w2}, Exit-End Flowing Stream Consisting of Water (fraction)	df_w/dS_w, Slope of Fractional Flow Curve	Q_i, PV of Cumulative Injected Water	$\bar{S}_w$, Average Water Saturation (fraction PV)
0.469	0.798	2.16	0.463	0.563
0.495	0.848	1.75	0.572	0.582
0.520	0.888	1.41	0.711	0.600
0.546	0.920	1.13	0.887	0.617
0.572	0.946	0.851	1.176	0.636
0.597	0.965	0.649	1.540	0.652
0.622	0.980	0.477	2.100	0.666
0.649	0.990	0.317	3.157	0.681
0.674	0.996	0.195	5.13	0.694
0.700	1.000	0.102	9.80	0.700

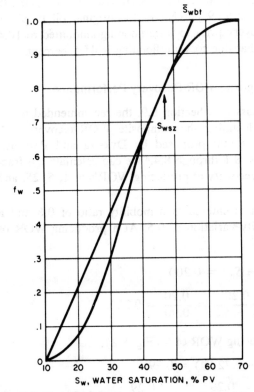

Fig. E.1 Fractional flow curve, example problem.

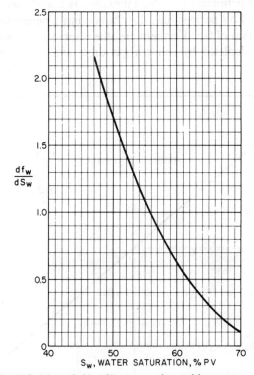

Fig. E.2 Plot of df_w/dS_w, example problem.

through, is found by

$$Q_i = \frac{1}{\left(\dfrac{df_w}{dS_w}\right)_{S_{w2}}} \qquad \ldots \ldots \ldots (E.2)$$

Using Welge's equation,[1]

$$\overline{S}_w = S_{w2} + Q_i f_{o2} , \qquad \ldots \ldots \ldots (E.3)$$

we can obtain the value of the average saturation $\overline{S}_w$ corresponding to each set of values of S_{w2}, Q_i, f_{w2}, and f_{o2}. (See Table E.3 for the values of Q_i for the example problem.)

E.2 Calculation of Mobility Ratio

The value of the water-oil mobility ratio is calculated from Eq. 4.1a, which reads

$$M = \frac{k_{rw}}{\mu_w} \frac{\mu_o}{k_{ro}} \qquad \ldots \ldots \ldots (E.4)$$

The relative permeability to water at the average water saturation (56.3 percent PV) is read from Fig. 2.17a as equal to 0.4. The relative permeability to oil ahead of the flood front is equal to 1.0. So

$$M = \frac{0.4}{0.5} \times \frac{1.0}{1.0} = 0.80.$$

E.3 Calculation of Ultimate Waterflood Recovery

Let us further consider that the core analysis from the example reservoir, when arranged from maximum values of permeability to the minimum, is that shown in Fig. E.3. This permeability distribution has a permeability variation of 0.5. Furthermore, the reservoir has been partially depleted by solution gas drive and the recovery to date has been 10.4 percent of the original oil in place.

To calculate the ultimate waterflood recovery to a producing water cut of 98 percent, we first go to Table E.3 and determine that this water cut corresponds to an average water saturation of 66.6 percent PV.

The initial stock-tank oil in place in a barrel of total pore volume (oil reservoir volume factor of 1.29 at the original reservoir pressure) is

$$\frac{S_{oi}}{B_{oi}} = \frac{1 - S_{wc}}{B_{oi}} = \frac{0.90}{1.29} = 0.698 \text{ STB}.$$

At a producing water cut of 98 percent, the stock-tank oil remaining in 1 bbl of total pore volume in the swept portion of the reservoir is

$$\frac{\overline{S}_o}{B_o} = \frac{1 - \overline{S}_w}{B_o} = \frac{0.334}{1.2} = 0.278 \text{ STB}.$$

The oil reservoir volume factor of 1.20 is that at the start of waterflooding. The oil remaining in the unswept portion, per barrel of total pore volume is

$$\frac{S_{oi}}{B_o} = \frac{1 - S_{wc}}{B_o} = \frac{1.0 - 0.1}{1.2} = 0.75 \text{ STB}.$$

Approximating the volumetric sweep efficiency by the term $(1 - V^2)/M$, we find this equals 0.9375. The total oil remaining in 1 barrel of total pore volume is

$$0.9375 \times 0.278 + (1 - 0.9375) \times 0.75 = 0.3075 \text{ STB}.$$

Thus, the total oil recovery is $(0.698 - 0.3075)/0.698 = 0.559$ or 55.9 percent of the original oil in place. Since recovery prior to waterflooding amounted to 10.4 percent, that due to waterflooding is 45.5 percent.

E.4 Composite WOR-Recovery Performance

As discussed in Section 8.9, the recommended method for calculating the composite WOR-recovery performance is that proposed by Dykstra and Parsons.[2] Using Figs. 8.1 through 8.4,[3] we can calculate the fractional oil recovery at producing WOR's of 1, 5, 25, and 100.

Fig. 8.1 is entered at a mobility ratio of 0.8 and a permeability variation of 0.5. At a producing WOR of 1, we read

$$E_R (1 - S_w) = 0.200$$

$$E_R = \frac{0.20}{1 - S_w} = \frac{0.20}{0.90} = 0.222.$$

At a producing WOR of 5 (Fig. 8.2),

$$E_R (1 - 0.72 S_w) = 0.29$$

$$E_R = \frac{0.29}{1 - 0.72 S_w} = \frac{0.29}{0.928} = 0.312.$$

At a producing WOR of 25 (Fig. 8.3),

$$E_R (1 - 0.52 S_w) = 0.38$$

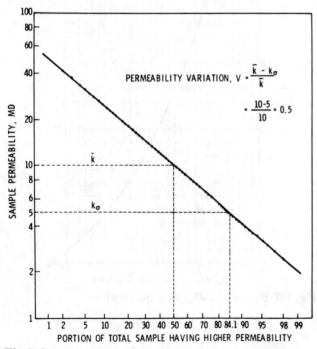

Fig. E.3 Permeability distribution, example problem.

APPENDIX E

$$E_R = \frac{0.38}{1. - 0.52\, S_w} = \frac{0.38}{0.948} = 0.400$$

At a producing WOR of 100 (Fig. 8.4),

$$E_R (1 - 0.40\, S_w) = 0.43$$

$$E_R = \frac{0.43}{1 - 0.40\, S_w} = \frac{0.43}{0.96} = 0.459.$$

Subtracting the current primary recovery, 10.4 percent of the initial in-place oil, from these figures yields the expected waterflood recoveries at these producing water cuts.

Fig. E.4 shows these calculated waterflood recoveries plotted vs producing WOR's. At a producing water cut of 98 percent, the producing WOR is 0.98/0.02 = 49.

Interpolating on Fig. E.4 at a WOR of 49 yields a waterflood recovery of 33.6 percent of the original oil in place. This compares with 45.5 percent recovery estimated in Section E.3. The difference is due to the assumption, inherent in the use of Figs. 8.1 through 8.4, that the reservoir has flow properties like those of California oil sands.

E.5 Composite Injection and Producing Rates, WOR, and Recovery vs Time

The recommended method to obtain waterflood performance with time for a five-spot pattern is (1) the calculational approach of Craig, Geffen, and Morse[4] for relating oil recovery and producing WOR to cumulative injected water, coupled with (2) the correlation of Caudle and Witte[5] for calculating five-spot water injection rates. As discussed in Chapter 8, the performance of a five-spot flood approximates that of many other patterns.

As a means of including the effects of reservoir nonuniformity, the performance of a stratified five-spot can be calculated. To determine the recommended minimum number of layers, refer to Table 7.1. This table shows that at a mobility ratio of 0.8 and a permeability variation of 0.5, 10 equal-thickness layers will yield the same calculated waterflood performance, above a WOR of 2.5, as many more layers. Table E.4 shows the permeabilities and thicknesses for 10 layers having a permeability variation of 0.5. These permeabilities are taken from Fig. E.3, representing average permeabilities for each 10-percent increment of the cumulative sample. These values are selected as the permeabilities at 5, 10, 15, . . . 95 percent of the cumulative sample.

The stratified reservoir in this example is composed of layers identical in all properties except permeability. Thus to calculate multilayer performance we can determine the performance of one layer (a so-called base layer), then obtain the performance in each of the remaining layers by adjusting the performance for the permeability contrast. The method of this adjustment will be discussed later in this appendix. For stratified reservoirs with layers differing in water-oil relative permeability characteristics, the performance of each layer must be calculated individually. In both cases the composite performance is the sum of the individual layer performances.

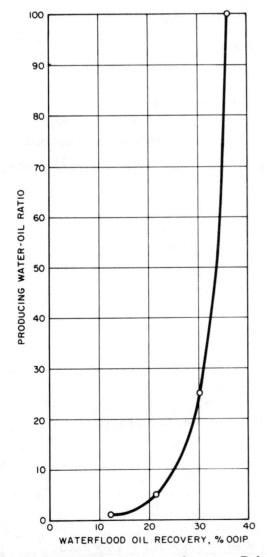

Fig. E.4 Predicted WOR-recovery performance, Dykstra-Parsons method.

TABLE E.4—PROPERTIES OF 10 EQUAL-THICKNESS LAYERS HAVING A PERMEABILITY VARIATION OF 0.5 FOR EXAMPLE RESERVOIR

Layer	Permeability (md)	Thickness (ft)
1	31.5	5.0
2	20.5	5.0
3	16.0	5.0
4	13.1	5.0
5	10.9	5.0
6	8.2	5.0
7	7.7	5.0
8	6.3	5.0
9	4.9	5.0
10	3.2	5.0

This method of predicting five-spot water injection performance is valid either with or without free gas present, provided that there is no trapped gas behind the flood front. The calculations are not valid for floods in which there is bottom water.

These calculations assume a vertical sweep efficiency of 100 percent in each layer—that is, the fluids are not segregated by gravity.

The performance of a waterflood can be divided into four stages.

Stage One is the period of radial flow out from the injectors from the start of injection until the oil banks, formed around adjacent injectors, meet. The meeting of adjacent oil banks is termed "interference".

Stage Two is the period from interference until fillup of the pre-existing gas space. Fillup is the start of oil production response.

Stage Three is the period from fillup to water breakthrough at the producing wells. Breakthrough marks the beginning of water production.

Stage Four is the period from water breakthrough to floodout.

In this appendix we shall discuss the calculation of waterflood performance during each of these stages.

To begin, several initial calculations are required. These are shown in Table E.5. These calculations include determination of (1) pore volume, (2) oil in place at the waterflood, and (3) areal sweep efficiency at water breakthrough. The other calculations shown on Table E.5 will be discussed later.

Performance Prior to Interference

During this stage of the flood, radial flow prevails. The water injection at interference, W_{ii}, is equal to the

TABLE E.5 — INITIAL CALCULATIONS, EXAMPLE PROBLEM

1. Pore volume $(V_P) = 7{,}758\, A\, h\, \phi = 7{,}758 \times 40 \times 5.0 \times 0.20 = 310{,}320$ bbl.

2. Stock-tank oil in place at start of waterflood $= \dfrac{V_P \times S_{oi}}{B_{oi}} = \dfrac{310{,}320 \times 0.75}{1.20} = 193{,}950$ bbl.

3. Areal sweep efficiency at water breakthrough, $E_{Abt} = 0.717$ from Fig. E.6 at $M = 0.8$.

4. Injected water at breakthrough, $W_{ibt} = V_P \times E_{Abt} \times (\overline{S}_{wbt} - S_{wc})$
   $= 310{,}320 \times 0.717 \times (0.563 - 0.10) = 103{,}020$ bbl.

   a. Maximum value of initial gas saturation at which prediction method is accurate, $S_{gi}^* = C\,(\overline{S}_{oi} - \overline{S}_{obt})$.

   b. The value of C is found from Fig. E.7 at $M = 0.8$. $S_{gi}^* = 1.18\,(0.75 - 0.437) = 0.369$ or 36.9 percent.

   c. Since the initial gas saturation, 15 percent PV, is less than the value of S_{gi}^*, the prediction method is applicable.

5. Injected water at interference, $W_{ii} = \dfrac{\pi h \phi\, S_{gi}\, r_{ei}^2}{5.61}$, where r_{ei} is half the distance between adjacent injectors, or 660 ft.

   $W_{ii} = \dfrac{\pi \times 5.0 \times 0.20 \times 0.15 \times (660)^2}{5.61} = 36{,}572$ bbl.

6. Injected water at fillup, $W_{if} = V_P \times S_{gi} = 310{,}320 \times 0.15 = 46{,}550$ bbl.

7. Areal sweep efficiency, E_A, at an injected water volume, W_i, prior to water breakthrough

   $E_A = \dfrac{W_i}{V_P\,(\overline{S}_{wbt} - S_{wc})} = \dfrac{W_i}{310{,}320\,(0.563 - 0.1)} = \dfrac{W_i}{143{,}678}.$

8. Outer radius of oil bank, r_e, prior to interference:

   $r_e = \left(\dfrac{5.61\, W_i}{\pi h \phi\, S_{gi}}\right)^{1/2} = \left(\dfrac{5.61 \times W_i}{\pi \times 5.0 \times 0.20 \times 0.15}\right)^{1/2} = \left(11.915\, W_i\right)^{1/2}.$

9. Outer radius of water flood front, r, prior to interference:

   $r = r_e \left(\dfrac{S_{gi}}{\overline{S}_{wbt} - S_{wc}}\right)^{1/2} = r_e \left(\dfrac{0.15}{0.563 - 0.10}\right)^{1/2} = 0.5692\, r_e.$

10. Water injection rate, i_w, to interference for layer of 31.5 md or 0.0315 darcies and a Δp of 3,000 psi:

    $i_w = 7.07 \times 10^{-3}\, hk\, \Delta p \left(\dfrac{\mu_w}{k_{rw}} \ln \dfrac{r}{r_w} + \dfrac{\mu_o}{k_{ro}} \ln \dfrac{r_e}{r}\right)^{-1}$

    $= 7.07 \times 10^{-3} \times 5.0 \times 31.5 \times 3{,}000 \left(\dfrac{\mu_w}{k_{rw}} \ln \dfrac{r}{r_w} + \dfrac{\mu_o}{k_{ro}} \ln \dfrac{r_e}{r}\right)^{-1}$

    $= 3{,}340 \left(\dfrac{\mu_w}{k_{rw}} \ln \dfrac{r}{r_w} + \dfrac{\mu_o}{k_{ro}} \ln \dfrac{r_e}{r_w}\right)^{-1}.$

11. Base water injection rate, $i_{base} = \dfrac{3.541 \times 10^{-3}\, h\, k\, k_{ro}\, \Delta p}{\mu_o\,(\ln d/r_w - 0.619)} = \dfrac{3.541 \times 10^{-3} \times 5.0 \times 31.5 \times 1.0 \times 3{,}000}{1.0 \left(\ln \dfrac{932}{1} - 0.619\right)} = 269.1$ B/D

APPENDIX E

free gas saturation in the cylindrical portion of the reservoir with radius r_{ei}. The value of r_{ei} is one-half the distance between adjacent injectors. From Table E.5, we find that the injected water volume at interference is 36,572 bbl.

The water injection rate prior to interference[6] is

$$i_w = \frac{7.07 \times 10^{-3} \, h \, k \, \Delta p}{\left(\frac{\mu_w}{k_{rw}} \ln \frac{r}{r_w} + \frac{\mu_o}{k_{ro}} \ln \frac{r_e}{r} \right)}, \quad \ldots \quad (E.5)$$

where

r_e = outer radius of the oil bank,

r = outer radius of the water flood front.

Table E.6 shows the calculation of the injection rate variation from the start of water injection to interference. Table E.6 utilizes some of the equations shown in Table E.5. This calculation shows that interference will occur at 75.3 days after the start of water injection.

Performance from Interference to Fillup

During this time, flow is not strictly radial. From Table E.6, the water injection rate at interference is known. It is possible to calculate the water injection rate at fillup as follows.

At fillup, the cumulative injected water, W_{if}, is 46,550 bbl from Table E.5. The calculation shown on this table utilizes the equation

$$W_{if} = S_{gi} \times V_P \quad \ldots \ldots \ldots \ldots \quad (E.6)$$

This equation assumes that while fillup is occurring, the oil producing rate is either zero or negligible, compared with the water injection rate. If the oil producing rate prior to fillup is significant, the water volume injected to fillup must be increased by the reservoir volume of the oil produced from the start of injection to fillup. This obviously also increases the time to fillup. It also causes the fillup time calculation to be iterative.

The areal sweep efficiency at fillup can be found from Eq. E.7, which applies at fillup and thereafter.

$$E_A = \frac{W_i}{V_P (\overline{S}_{wbt} - S_{wc})} \quad \ldots \ldots \ldots \quad (E.7)$$

Substituting the value of W_i at fillup (46,550 bbl) and other values into Eq. E.7, we find that the areal sweep efficiency at fillup is 0.324, or 32.4 percent. Entering

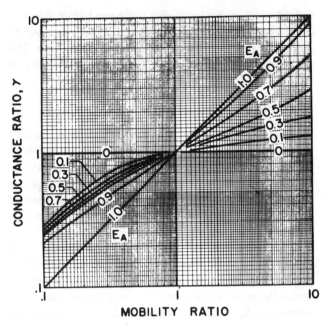

Fig. E.5 Conductance ratio, five-spot pattern.

Fig. E.5 at E_A of 0.324 and M of 0.8, the value of γ, the conductance ratio, is 0.96. The water injection rate at fillup and thereafter is

$$i_w = \gamma \times i_{\text{base}} \quad \ldots \ldots \ldots \ldots \quad (E.8)$$

This equation in effect defines the term γ, as used by Caudle and Witte.[5] The term i_{base} is equivalent to the injection rate of fluid having the same mobility as the reservoir oil in a liquid-filled pattern.

For the example problem, the injection rate at fillup is

$$i_{wf} = 0.96 \times 269.1 \text{ B/D} = 258.3 \text{ B/D},$$

where the value of i_{base} is obtained from Table E.5.

The incremental time occurring from interference to fillup is

$$\Delta t = \frac{W_{if} - W_{ii}}{0.5(i_{wi} + i_{wf})} = \frac{46{,}550 - 36{,}560}{0.5(418.9 + 258.3)}$$

$$= 29.5 \text{ days}.$$

Thus, the time to fillup is $80.00 + 29.50 = 109.50$ days.

Performance from Fillup to Breakthrough

Table E.7 shows the calculated water injection rate

TABLE E.6 — PERFORMANCE PRIOR TO INTERFERENCE, EXAMPLE FIVE-SPOT WATERFLOOD PROBLEM

(1) W_i (assume)	(2) r_e^2	(3) r_e $(2)^{1/2}$	(4) r	(5) $\frac{\mu_w}{k_{rw}} \ln \frac{r}{r_w}$	(6) $\frac{\mu_o}{k_{ro}} \ln \frac{r_e}{r}$	(7) (5) + (6)	(8) i_w (BWPD)	(9) $i_{w,\text{avg}}$	(10) $\Delta t = \Delta W_i / i_{w,\text{avg}}$	(11) $t = \Sigma(\Delta t)$ (days)
0	0	0	0							
500	5958	77.2	43.9	4.7274	0.5635	5.2909	631.3	631.3	0.79	0.79
5000	59575	244.1	138.9	6.1672	0.5635	6.7307	496.2	563.8	7.98	8.77
10000	119150	345.2	196.5	6.6008	0.5635	7.1643	466.2	481.2	10.39	19.16
15000	178725	422.8	240.7	6.8544	0.5635	7.4179	450.3	458.3	10.91	30.07
20000	238300	488.2	277.9	7.0341	0.5635	7.5976	439.6	445.0	11.24	41.31
25000	297875	545.8	310.7	7.1735	0.5635	7.7370	431.7	435.7	11.48	52.79
30000	357450	597.9	340.3	7.2873	0.5635	7.8508	425.5	428.6	11.67	64.46
35000	417025	645.8	367.6	7.3837	0.5635	7.9472	420.3	422.9	11.82	76.28
36560	435600	660.0	375.7	7.4110	0.5635	7.9745	418.9	419.6	3.72	80.00

and oil producing rate during the period from fillup to water breakthrough. During this period the *reservoir* oil producing rate is equal to the water injection rate. The calculations shown in Table E.7 use this equality as well as Eqs. E.7 and E.8.

Performance After Water Breakthrough

After water breakthrough, the WOR is computed on the basis of the amounts of water and oil flowing from the swept region (from frontal drive theory) and the oil displaced as the swept area increases. The oil displaced from the newly swept portion of the pattern is assumed to be that displaced by the water saturation just behind the stabilized zone, S_{wsz}.

The incremental oil produced from the newly swept region, $\triangle N_{pu}$, will then be the incremental increase in areal sweep efficiency multiplied by the difference between the water saturation, S_{wsz}, and the connate water saturation, S_{wc}, and also multiplied by the pore volume.

$$\triangle N_{pu} = \triangle E_A (S_{wsz} - S_{wc})(V_P) \quad \ldots \quad (E.9)$$

The term $\left(\dfrac{\triangle E_A}{\triangle W_i/W_{ibt}}\right)$ is a convenient grouping.

Let us rearrange Eq. E.9 to yield

$$\triangle N_{pu} = \left(\frac{\triangle E_A}{\triangle W_i/W_{ibt}}\right)(S_{wsz} - S_{wc})(V_P)\frac{\triangle W_i}{W_{ibt}} \quad \ldots \quad (E.10)$$

After fillup, the water injection rate equals the total production rate.

If $\triangle W_i$ is set equal to 1 bbl of total production (both oil and water), the oil produced from the newly swept region, in barrels, is

$$\triangle N_{pu} = \left(\frac{\triangle E_A}{\triangle W_i/W_{ibt}}\right)(S_{wsz} - S_{wc})(V_P)\frac{1}{W_{ibt}} \quad \ldots \quad (E.11)$$

The term W_{ibt} is the volume of water injected to water breakthrough. This is equal to the pore volume multiplied by the areal sweep efficiency at breakthrough, multiplied by the difference between the average water saturation in the swept region at breakthrough, $\overline{S}_{wbt}$, and the connate water saturation.

$$W_{ibt} = (V_P)(E_{Abt})(\overline{S}_{wbt} - S_{wc}) \quad \ldots \quad (E.12)$$

Substituting W_{ibt} in the previous equation, we obtain

$$\triangle N_{pu} = \lambda \frac{S_{wsz} - S_{wc}}{E_{Abt}(\overline{S}_{wbt} - S_{wc})}, \quad \ldots \quad (E.13)$$

where

$$\lambda = \frac{\triangle E_A}{\triangle W_i/W_{ibt}} \quad \ldots \quad (E.14)$$

The incremental oil produced from the previously swept region, $\triangle N_{ps}$, for 1 bbl of total production is

$$\triangle N_{ps} = f_{o2}(1 - \triangle N_{pu}) \quad \ldots \quad (E.15)$$

The incremental water produced from the swept region, $\triangle W_{ps}$, per 1 bbl of total production is

$$\triangle W_{ps} = 1 - (\triangle N_{ps} + \triangle N_{pu}) \quad \ldots \quad (E.16)$$

This yields a producing WOR, at reservoir pressure, of

$$WOR_p = \frac{1 - (\triangle N_{ps} + \triangle N_{pu})}{\triangle N_{ps} + \triangle N_{pu}}, \quad \ldots \quad (E.17)$$

or an atmospheric producing WOR of

$$WOR = WOR_p \times B_o \quad \ldots \quad (E.18)$$

This equation assumes that there is no change in water volume from reservoir to surface conditions.

Water Injection with No Free Gas Initially Present

A step-by-step worksheet for five-spot water injection calculations is shown in Table E.8.

To start the calculation, select a value of W_i (Column 1). Dividing Column 1 by W_{ibt} (from Eq. E.12) yields the value of W_i/W_{ibt} (Column 2). On Fig. E.7, plot the value of E_{Abt} at $W_i/W_{ibt} = 1.00$. A straight line (shown dashed on Fig. E.8) is drawn through this point parallel to the other straight lines.

The relationship shown on Fig. E.8 can be written as

$$E_A = E_{Abt} + 0.2749 \ln W_i/W_{ibt} \quad \ldots \quad (E.19)$$

An alternative to the use of Fig. E.8 to find the value of E_A is to use Eq. E.19. Of course, the value of E_A cannot exceed 1.0.

At areal sweep efficiencies below 1.0 (100 percent), the value of Q_i/Q_{ibt} (Column 4) is read from Table E.9 for the particular value of E_{Abt} and the value of W_i/W_{ibt} (Column 2) using interpolation if necessary. The values shown in Table E.9 are taken from the

TABLE E.7 — PERFORMANCE FROM FILLUP TO WATER BREAKTHROUGH, EXAMPLE FIVE-SPOT WATERFLOOD PROBLEM

(1) W_i (assume)	(2) E_A (fraction)	(3) γ From Fig. E.5	(4) $i_w = i_{base} \times (3)$	(5) $i_{w,avg}$	(6) $\triangle t = \triangle W_i/i_{w,avg}$	(7) $t = \Sigma(\triangle t)$ (days)	(8) q_o (4) ÷ B_o	(9) $W_i - W_{if}$ (1) − W_{if}	(10) N_p (9) ÷ B_o (bbl)	(11) N_p (10) ÷ OIP (% oil in place)
46550	0.324	0.96	258.6			109.50	215.2	0	0	0
50000	0.348	0.95	255.6	257.1	13.42	122.92	213.0	3450	2875	1.48
60000	0.418	0.94	253.0	254.3	39.32	162.24	210.8	13450	11208	5.78
70000	0.487	0.94	253.0	253.0	39.53	201.77	210.8	23450	19542	10.08
80000	0.557	0.93	250.3	251.6	39.75	241.52	208.6	33450	27875	14.37
90000	0.626	0.92	247.6	248.9	40.18	281.70	206.3	43450	36208	18.67
100000	0.696	0.92	247.6	247.6	40.39	322.09	206.3	53450	44542	22.97
103020	0.717	0.91	244.9	246.2	12.27	334.36	204.1	56470	47058	24.26

APPENDIX E

relationship

$$\frac{Q_i}{Q_{ibt}} = 1 + E_{Abt} \int_1^{W_i/W_{ibt}} \frac{1}{E_A} d(W_i/W_{ibt}).$$

At any step in the calculations when the value of E_A (Column 3) is unity, the incremental increase in Q_i/Q_{ibt} (Column 4) is equal to the incremental increase in W_i/W_{ibt} (Column 2) multiplied by E_{Abt}.

The value of Q_i/Q_{ibt} (Column 4) is read from Table E.9 for the particular value of E_{Abt} and the value of W_i/W_{ibt} (Column 2), using interpolation if necessary. The value of the term Q_i/Q_{ibt} is different from that of W_i/W_{ibt} because the Q_i term is based upon the *water-contacted* pore volume, which increases with time as the invaded area increases.

The term Q_{ibt} is the injected water volume at water breakthrough, expressed in water-contacted PV's. This is equal to the difference between the average water saturation of the water-contacted region at water breakthrough, $\overline{S}_{wbt}$, and the connate water saturation, S_{wc}.

Q_i, the cumulative injected water volume, in water-contacted PV's (Column 5), is obtained by multiplying Column 4 by Q_{ibt}.

The term $(df_w/dS_w)_{S_{w2}}$ (Column 6) is the reciprocal of Column 5 (see Eq. E.2). The value of S_{w2}, the water saturation at the producing wells (Column 7), is read from the plot of df_w/dS_w vs S_w (Fig. E.2) as the water saturation having a value of df_w/dS_w equal to that given in Column 6.

The term f_{o2}, the oil fraction of the fluid flowing from the water-contacted region (Column 8), is equal to 1 minus the value of f_w corresponding to the saturation, S_{w2}. The value of f_w is found from Fig. E.1.

TABLE E.8 — PERFORMANCE AFTER WATER BREAKTHROUGH, EXAMPLE FIVE-SPOT WATERFLOOD PROBLEM

(1)	(2)	(3)	(4)	(5)	(6)	(7)	(8)	(9)
W_i	W_i/W_{ibt} (1) ÷ W_{ibt}	E_A From Fig. E.8 or Eq. E.19	Q_i/Q_{ibt} From Table E.9	Q_i (4) × Q_{ibt}	$\left(\frac{df_w}{dS_w}\right)_{S_{w2}}$	S_{w2} From Fig. E.2	f_{o2} From Fig. E.1	$\overline{S}_w$ in swept area (7) + [(5) × (8)]
103020	1.0	0.717	1.000	0.463	2.159	0.469	0.200	0.563
123620	1.2	0.767	1.193	0.552	1.810	0.492	0.157	0.579
144230	1.4	0.809	1.375	0.636	1.570	0.507	0.130	0.590
164830	1.6	0.846	1.548	0.717	1.394	0.524	0.107	0.601
185440	1.8	0.879	1.715	0.794	1.259	0.534	0.095	0.610
206040	2.0	0.906	1.875	0.869	1.151	0.543	0.080	0.613
257550	2.5	0.969	2.256	1.046	0.956	0.562	0.063	0.628
309060	3.0	1.000	2.619	1.214	0.823	0.575	0.051	0.637
412080	4.0	1.000	3.336	1.545	0.647	0.597	0.037	0.653
515100	5.0	1.000	4.053	1.877	0.533	0.611	0.027	0.660
618120	6.0	1.000	4.770	2.208	0.453	0.622	0.020	0.664
824160	8.0	1.000	6.204	2.872	0.348	0.637	0.015	0.676
1030200	10.0	1.000	7.638	3.536	0.283	0.650	0.010	0.683
1545300	15.0	1.000	11.223	5.196	0.192	0.672	0.005	0.697

(10)	(11)	(12)	(13)	(14)	(15)	(16)	(17)
λ From Eq. E.20	ΔN_{pu} From Eq. E.13	$1 - \Delta N_{pu}$ 1.0 − (11)	ΔN_{ps} (8) × (12)	$\Delta N_{pu} + \Delta N_{ps}$ (11) + (13)	WOR_v [1.0−(14)] ÷ (14)	WOR (15) × B_o	$\overline{S}_w - S_{wc}$ in Swept Area (9) − S_{wc}
0.2749	0.3056	0.6944	0.1388	0.4444	1.25	1.50	0.463
0.2291	0.2545	0.7455	0.1170	0.3715	1.69	2.03	0.479
0.1963	0.2182	0.7818	0.1016	0.3198	2.13	2.55	0.490
0.1718	0.1910	0.8090	0.0866	0.2776	2.60	3.12	0.501
0.1527	0.1697	0.8303	0.0789	0.2486	3.02	3.63	0.510
0.1375	0.1528	0.8472	0.0678	0.2206	3.53	4.24	0.513
0.1100	0.1223	0.8777	0.0552	0.1775	4.63	5.56	0.528
0.000	0.0000	1.0	0.0510	0.0510	18.61	22.3	0.537
0.000	0.0000	1.0	0.0370	0.0370	26.03	31.2	0.553
0.000	0.0000	1.0	0.0270	0.0270	36.0	43.2	0.560
0.000	0.0000	1.0	0.0200	0.0200	49.0	58.8	0.564
0.000	0.0000	1.0	0.0150	0.0150	65.7	78.8	0.576
0.000	0.0000	1.0	0.0100	0.0100	99.0	118.8	0.583
0.000	0.0000	1.0	0.0050	0.0050	199.0	238.8	0.597

(18)	(19)	(20)	(21)	(22)	(23)	(24)	(25)	(26)	(27)	(28)	(29)
[(3) × (17)] − S_{gi} (fraction PV)	OIL RECOVERY (18) ÷ S_{oi} (fraction OIP)	(19) × STOIP (bbl)	k_{rw} at S_w	M From Eq. E.4	γ From Fig. E.5	i_w (23) × i_{base}	$i_{w, avg}$	ΔW_i	Δt (26) ÷ (25)	$t = \Sigma(\Delta t)$ (days)	q_o (14) × (24) ÷ B_o
0.1820	0.2426	47,058	0.400	0.800	0.91	244.9	248.9	20600	82.75	334.36	90.7
0.2174	0.2899	56,218	0.430	0.860	0.94	252.9	255.6	20610	80.62	417.12	78.3
0.2464	0.3286	63,722	0.450	0.900	0.96	258.3	261.0	20600	78.92	497.74	68.8
0.2738	0.3651	70,817	0.480	0.960	0.98	263.7	266.4	20610	77.36	576.66	61.0
0.2983	0.3977	77,138	0.500	1.000	1.00	269.1	271.8	20600	75.79	654.02	56.7
0.3153	0.4204	81,534	0.510	1.020	1.02	274.5	282.6	51510	182.30	729.81	50.5
0.3616	0.4822	93,518	0.542	1.084	1.08	290.6	296.0	51510	174.01	912.11	43.0
0.3870	0.5160	100,078	0.560	1.120	1.12	301.4	312.2	103020	330.03	1086.13	12.8
0.4030	0.5373	104,216	0.600	1.200	1.20	322.9	329.6	103020	312.52	1416.16	10.0
0.4100	0.5467	106,026	0.625	1.250	1.25	336.4	339.1	103020	303.83	1728.67	7.6
0.4140	0.5520	107,060	0.635	1.270	1.27	341.8	351.2	206040	586.72	2032.51	5.7
0.4260	0.5680	110,164	0.670	1.340	1.34	360.6	366.0	206040	562.99	2619.22	4.5
0.4330	0.5773	111,974	0.690	1.380	1.38	371.4	379.4	515100	1357.56	3182.21	3.1
0.4470	0.5960	115,594	0.720	1.440	1.44	387.5				4539.77	1.6

Column 9, the average water saturation in the swept area, is found from Eq. E.3.

Using Eqs. E.14 and E.19, we can derive an analytical expression for the term λ. This expression is

$$\lambda = 0.2749 \, (W_i/W_{ibt})^{-1} \quad \ldots \ldots \quad (E.20)$$

At any step in the calculations when the value of E_A (Column 3) is unity, the value of λ is zero.

The incremental oil produced from the newly water-contacted pattern region per barrel of produced fluid, $\triangle N_{pu}$ (Column 11), is calculated by Eq. E.13.

Column 12 is 1 minus Column 11.

The incremental oil produced from the swept region per barrel of produced fluid, $\triangle N_{ps}$ (Column 13), is calculated from Eq. E.15 by multiplying Column 8 by Column 12.

The flowing WOR in bbl/bbl at reservoir pressure (Column 15) is calculated from Eq. E.17 by dividing 1 minus Column 14 by Column 14.

The flowing WOR in bbl/bbl at atmospheric pressure

TABLE E.9 — VALUES OF Q_i/Q_{ibt} FOR VARIOUS VALUES OF BREAKTHROUGH AREAL SWEEP EFFICIENCY

W_i/W_{ibt}	E_{Abt}, percent									
	50.	51.	52.	53.	54.	55.	56.	57.	58.	59.
1.0	1.000	1.000	1.000	1.000	1.000	1.000	1.000	1.000	1.000	1.000
1.2	1.190	1.191	1.191	1.191	1.191	1.191	1.191	1.191	1.192	1.192
1.4	1.365	1.366	1.366	1.367	1.368	1.368	1.369	1.369	1.370	1.370
1.6	1.529	1.530	1.531	1.532	1.533	1.535	1.536	1.536	1.537	1.538
1.8	1.684	1.686	1.688	1.689	1.691	1.693	1.694	1.696	1.697	1.699
2.0	1.832	1.834	1.837	1.839	1.842	1.844	1.846	1.849	1.851	1.853
2.2	1.974	1.977	1.981	1.984	1.987	1.990	1.993	1.996	1.999	2.001
2.4	2.111	2.115	2.119	2.124	2.127	2.131	2.135	2.139	2.142	2.146
2.6	2.244	2.249	2.254	2.259	2.264	2.268	2.273	2.277	2.282	2.286
2.8	2.373	2.379	2.385	2.391	2.397	2.402	2.407	2.413	2.418	2.422
3.0	2.500	2.507	2.513	2.520	2.526	2.533	2.539	2.545	2.551	2.556
3.2	2.623	2.631	2.639	2.646	2.653	2.660	2.667	2.674	2.681	2.687
3.4	2.744	2.752	2.761	2.770	2.778	2.786	2.793	2.801	2.808	2.816
3.6	2.862	2.872	2.881	2.891	2.900	2.909	2.917	2.926	2.934	2.942
3.8	2.978	2.989	3.000	3.010	3.020	3.030	3.039	3.048	3.057	3.066
4.0	3.093	3.105	3.116	3.127	3.138	3.149	3.159	3.169	3.179	3.189
4.2	3.205	3.218	3.231	3.243	3.254	3.266	3.277	3.288	3.299	3.309
4.4	3.316	3.330	3.343	3.357	3.369	3.382	3.394	3.406	3.417	3.428
4.6	3.426	3.441	3.455	3.469	3.483	3.496	3.509	3.521	3.534	3.546
4.8	3.534	3.550	3.565	3.580	3.594	3.609	3.622	3.636	3.649	
5.0	3.641	3.657	3.674	3.689	3.705	3.720	3.735			
5.2	3.746	3.764	3.781	3.798	3.814	3.830				
5.4	3.851	3.869	3.887	3.905	3.922					
5.6	3.954	3.973	3.993	4.011						
5.8	4.056	4.077	4.097							
6.0	4.157	4.179								
6.2	4.257									
Values of W_i/W_{ibt} at which $E_A = 100$ percent										
	6.164	5.944	5.732	5.527	5.330	5.139	4.956	4.779	4.608	4.443

W_i/W_{ibt}	E_{Abt}, percent									
	60.	61.	62.	63.	64.	65.	66.	67.	68.	69.
1.0	1.000	1.000	1.000	1.000	1.000	1.000	1.000	1.000	1.000	1.000
1.2	1.192	1.192	1.192	1.192	1.192	1.192	1.193	1.193	1.193	1.193
1.4	1.371	1.371	1.371	1.372	1.372	1.373	1.373	1.373	1.374	1.374
1.6	1.539	1.540	1.541	1.542	1.543	1.543	1.544	1.545	1.546	1.546
1.8	1.700	1.702	1.703	1.704	1.706	1.707	1.708	1.709	1.710	1.711
2.0	1.855	1.857	1.859	1.861	1.862	1.864	1.866	1.868	1.869	1.871
2.2	2.004	2.007	2.009	2.012	2.014	2.016	2.019	2.021	2.023	2.025
2.4	2.149	2.152	2.155	2.158	2.161	2.164	2.167	2.170	2.173	2.175
2.6	2.290	2.294	2.298	2.301	2.305	2.308	2.312	2.315	2.319	2.322
2.8	2.427	2.432	2.436	2.441	2.445	2.449	2.453	2.457	2.461	2.465
3.0	2.562	2.567	2.572	2.577	2.582	2.587	2.592	2.597	2.601	2.606
3.2	2.693	2.700	2.705	2.711	2.717	2.723	2.728	2.733	2.738	2.744
3.4	2.823	2.830	2.836	2.843	2.849	2.855	2.862	2.867	2.873	
3.6	2.950	2.957	2.965	2.972	2.979	2.986	2.993			
3.8	3.075	3.083	3.091	3.099	3.107					
4.0	3.198	3.207	3.216	3.225						
4.2	3.319	3.329								
4.4	3.439									
Values of W_i/W_{ibt} at which $E_A = 100$ percent										
	4.285	4.132	3.984	3.842	3.704	3.572	3.444	3.321	3.203	3.088

APPENDIX E

(Column 18) is calculated from Eq. E.18 by multiplying Column 15 by the reservoir volume factor, B_o.

Column 17 is the difference between the average water saturation in the swept region (Column 9) and the connate water saturation.

The oil recovery in the entire five-spot pattern (Column 18), expressed as a fraction of the total pore volume, is calculated by multiplying Column 3 by Column 17.

The oil recovery in the entire five-spot pattern (Column 19), expressed as a fraction of the stock-tank oil initially in place, is calculated by dividing Column 18 by the initial oil saturation.

The oil recovery, in barrels, from the five-spot pattern under consideration (Column 20) is found by multiplying the stock-tank oil volume initially in place in the five-spot pattern by Column 19.

Column 21 is the relative permeability to water at the average water saturation (Column 9). Using this value of water relative permeability, we calculate the mobility ratio, Column 22, from Eq. E.4. As discussed in Chapter 4, the mobility ratio is constant up to breakthrough, but it increases after breakthrough as the average water saturation in the swept area increases.

Column 23 is γ, Caudle and Witte's conductivity ratio (Fig. E.5). The conductivity ratio is the ratio of the liquid-filled injectivity at any sweep and mobility ratio to the injectivity to oil in an initially liquid-filled pattern.

The injection rate to oil is calculated at the pressure difference between injector and producer to be used in the waterflood. Multiplying this rate by γ yields the value of the water injection rate (Column 24).

Column 25 is the average injection rate from the previous line to the current line. Column 26 is the incremental water injection volume over that same increment. Dividing Column 24 by Column 23 gives the incremental time from the previous to the present line in Table E.8 (Column 27). Summing these incremental

TABLE E.9 (CONTINUED) — VALUES OF Q_i/Q_{ibt} FOR VARIOUS VALUES OF BREAKTHROUGH AREAL SWEEP EFFICIENCY

W_i/W_{ibt}	E_{Abt}, percent									
	70.	71.	72.	73.	74.	75.	76.	77.	78.	79.
1.0	1.000	1.000	1.000	1.000	1.000	1.000	1.000	1.000	1.000	1.000
1.2	1.193	1.193	1.193	1.193	1.193	1.193	1.193	1.194	1.194	1.194
1.4	1.374	1.375	1.375	1.375	1.376	1.376	1.376	1.377	1.377	1.377
1.6	1.547	1.548	1.548	1.549	1.550	1.550	1.551	1.551	1.552	1.552
1.8	1.713	1.714	1.715	1.716	1.717	1.718	1.719	1.720	1.720	1.721
2.0	1.872	1.874	1.875	1.877	1.878	1.880	1.881	1.882	1.884	1.885
2.2	2.027	2.029	2.031	2.033	2.035	2.037	2.039	2.040	2.042	2.044
2.4	2.178	2.180	2.183	2.185	2.188	2.190	2.192	2.195	2.197	
2.6	2.325	2.328	2.331	2.334	2.337	2.340				
2.8	2.469	2.473	2.476	2.480						
3.0	2.610	2.614								
Values of W_i/W_{ibt} at which $E_A=100$ percent										
	2.978	2.872	2.769	2.670	2.575	2.483	2.394	2.309	2.226	2.147

W_i/W_{ibt}	E_{Abt}, percent									
	80.	81.	82.	83.	84.	85.	86.	87.	88.	89.
1.0	1.000	1.000	1.000	1.000	1.000	1.000	1.000	1.000	1.000	1.000
1.2	1.194	1.194	1.194	1.194	1.194	1.194	1.194	1.194	1.194	1.194
1.4	1.377	1.378	1.378	1.378	1.378	1.379	1.379	1.379	1.379	1.379
1.6	1.553	1.553	1.554	1.554	1.555	1.555	1.555	1.556	1.556	1.557	1.557
1.8	1.722	1.723	1.724	1.725	1.725	1.726	1.727	1.728		
2.0	1.886	1.887	1.888	1.890						
2.2	2.045									
Values of W_i/W_{ibt} at which $E_A=100$ percent										
	2.070	1.996	1.925	1.856	1.790	1.726	1.664	1.605	1.547	1.492

W_i/W_{ibt}	E_{Abt}, percent									
	90.	91.	92.	93.	94.	95.	96.	97.	98.	99.
1.0	1.000	1.000	1.000	1.000	1.000	1.000	1.000	1.000	1.000	1.000
1.2	1.194	1.195	1.195	1.195	1.195	1.195	1.195	1.195	1.195	1.195
1.4	1.380	1.380	1.380	1.380	1.381					
1.6	1.558									
Values of W_i/W_{ibt} at which $E_A=100$ percent										
	1.439	1.387	1.338	1.290	1.244	1.199	1.157	1.115	1.075	1.037

times yields the cumulative elapsed time (Column 28). Column 29, the oil producing rate, is equal to the water injection rate multiplied by $\frac{1}{B_o(1 + \text{WOR})}$, or Column 14 multiplied by Column 24, then divided by B_o.

Water Injection with Free Gas Initially Present

The method of calculation outlined previously is valid, with modification, for cases in which the initial free gas saturation is below a maximum value. If fillup is achieved at or before the stage of the flood at which the water flood front in a liquid-filled five-spot would start to cusp, the areal-sweep:WOR:injected-water-volume relationship of the five-spot with free gas present will be the same as that of an initially liquid-filled pattern. The maximum gas saturation at which this can occur depends upon the mobility ratio and the difference between the initial average oil saturation and the average oil saturation at water breakthrough in the swept region if no free gas were present. This maximum initial gas saturation S_{gi}^* is calculated as follows:

$$S_{gi}^* = C(\overline{S}_{oi} - \overline{S}_{obt}), \quad \ldots \quad (E.21)$$

where C is an experimentally determined function of the mobility ratio. Its value is read from Fig. E.7.

When the actual initial gas saturation is above the calculated maximum value, this method of calculation will yield WOR and recovery values at any injected water value higher than those actually expected.

The first step in the calculations is to compute from Eq. E.21 and Fig. E.7 the maximum initial gas saturation for which these water injection predictions are valid. This is to determine whether the calculated results will be realistic or conservative. S_{gi}^* is calculated on Table E.5 to be 36.9 percent PV for the example calculation. Thus these example calculations at a 15 percent initial gas saturation are valid.

The step-by-step guide for waterfloods in which initial gas saturations are present is the same as that for no initial gas.

The net effect of an initial gas saturation for a system in which there is no trapped gas is to reduce the oil recovery to any point by the volume of water injected to liquid fill-up.

Fig. E.9 shows the calculated WOR-recovery performance of a single layer. The sharp increase in producing WOR at about 49 percent recovery is at the point where the areal sweep efficiency becomes 100 percent.

Performance of Remaining Layers

To compute the performance of other layers that may differ in thickness, porosity, and permeability, the following logic is used. If subscript 1 denotes the first layer and subscript n denotes the nth layer,

$$t_n = t_1 \frac{k_1}{k_n} \frac{\phi_n}{\phi_1}, \quad \ldots \quad (E.22)$$

where t_n is the time to inject the same amount of water in PV's in the nth layer. At that time the oil producing rate, q_{on}, from the nth layer is found as follows:

$$q_{on} = q_{o1} \frac{k_n}{k_1} \frac{h_n}{h_1}; \quad \ldots \quad (E.23)$$

and the water injection rate, as follows:

$$i_{wn} = i_{wi} \frac{k_n}{k_1} \frac{h_n}{h_1} \quad \ldots \quad (E.24)$$

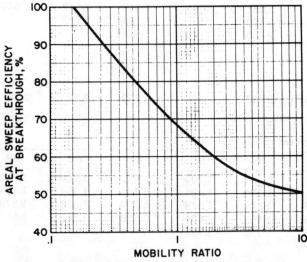

Fig. E.6 Areal sweep efficiency at water breakthrough, five-spot pattern.

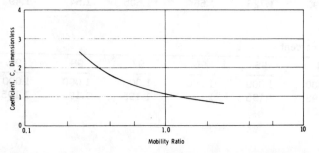

Fig. E.7 Relation between coefficient, C, and mobility ratio.

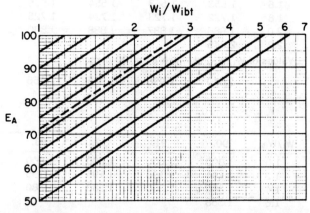

Fig. E.8 Effect of injected fluid volume on increase in areal sweep after water breakthrough.

APPENDIX E

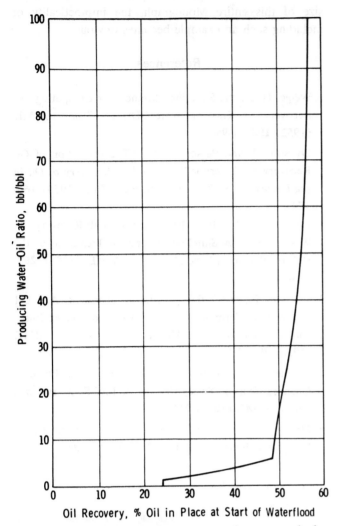

Fig. E.9 Calculated WOR-recovery performance, single-layer five-spot waterflood.

The areal sweep, recovery, and producing WOR are dependent only upon cumulative injected water volume, *expressed in PV's*.

To obtain composite performance, the water injection rates and oil producing rates and recoveries from each layer are computed at specified times. The injection rates, producing rates, and recoveries are summed. From the water and oil producing rates the composite WOR is calculated.

Fig. E.10 shows the calculated total water injection and oil production rate-time relationship for the example stratified waterflood problem, and Fig. E.11 shows the corresponding composite WOR-recovery performance.

The above calculation is only valid for reservoirs having (a) equal initial gas saturation in each layer and (b) insignificant producing rate prior to fillup. Fig. E.11 shows that the oil recovery by waterflooding, at a WOR of 49 (98 percent water cut), is about 55 percent of the oil in place at start of waterflooding. This recovery represents

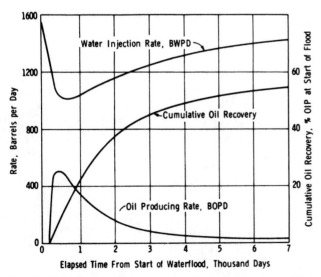

Fig. E.10 Calculated waterflood performance, stratified five-spot waterflood.

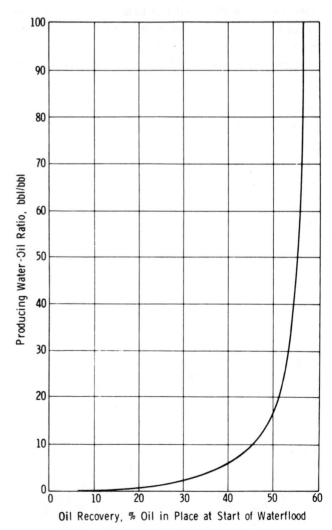

Fig. E.11 Calculated WOR-recovery performance, stratified five-spot waterflood.

$$0.55 \left[(1 - S_{gi} - S_{wc}) \frac{V_p}{B_o} \right]$$

$$= 0.55 \left[0.75 \frac{V_p}{1.2} \right] = 0.34375 \, V_p.$$

The original oil in place equals

$$(1 - S_{wc}) \frac{V_p}{B_{oi}} = 0.9 \frac{V_p}{1.29} = 0.6977 \, V_p.$$

Therefore, the waterflood recovery as a fraction of the original oil in place is

$$0.34375 \, V_p / 0.6977 \, V_p = 0.4926.$$

This value is comparable with but slightly higher than the value of 47.0 percent calculated for ultimate recovery in Section E.3. It is also higher than the value of 33.6 percent calculated by the Dykstra-Parsons approach.

E.6 Individual Well Performance

No attempt will be made in this Appendix to describe the calculation of individual well performance in a waterflood with dispersed injection wells. This type of calculation requires a computer program. Considering that documentation and user instructions for some of these programs will often fill one or more volumes the size of this entire Monograph, the impracticality of including such an example becomes obvious.

References

1. Welge, H. J.: "A Simplified Method for Computing Oil Recovery by Gas or Water Drive", *Trans.*, AIME (1952) **195**, 91-98.

2. Dykstra, H. and Parson, R. L.: "The Prediction of Oil Recovery by Waterflood", *Secondary Recovery of Oil in the United States,* 2nd ed., API, New York (1950) 160-174.

3. Johnson, C. E., Jr.: "Prediction of Oil Recovery by Water Flood—A Simplified Graphical Treatment of the Dykstra-Parsons Method", *Trans.*, AIME (1956) **207**, 345-346.

4. Craig, F. F., Jr., Geffen, T. M. and Morse, R. A.: "Oil Recovery Performance of Pattern Gas or Water Injection Operations from Model Tests", *Trans.*, AIME (1955) **204**, 7-15.

5. Caudle, B. H. and Witte, M. D.: "Production Potential Changes During Sweep-Out in a Five-Spot System", *Trans.*, AIME (1959) **216**, 446-448.

6. Muskat, M.: *Physical Principles of Oil Production*, McGraw-Hill Book Co., Inc., New York (1949) 682.

Nomenclature

a = distance between like wells in a row
A = area, sq ft
B_o = oil formation volume factor, volume at reservoir conditions divided by volume at standard conditions, dimensionless
B_g = gas formation volume factor, volume at reservoir conditions divided by volume at standard conditions, dimensionless
C = constant
d = distance between adjacent rows of wells
E_A = model areal sweep efficiency, area swept divided by total pattern area, fraction
E_D = microscopic oil displacement efficiency, volume of oil displaced divided by total oil volume, fraction
E_I = invasion or vertical sweep efficiency, fraction of vertical reservoir section contacted by the injected fluid
E_P = pattern areal sweep efficiency, pattern swept divided by total pattern area, fraction
E_R = recovery efficiency, fraction of initial oil in place recovered
E_V = volumetric sweep efficiency, fraction of total reservoir volume contacted by the injected fluid
f_{icw} = fraction of fluid produced at corner well that is injected fluid
f_{isw} = fraction of fluid produced at side well that is injected fluid
f_o = fraction of total flowing stream composed of oil
f_w = fraction of total flowing stream composed of water
F = function
F_{wo} = instantaneous producing WOR
g = acceleration due to gravity, ft/sq sec
h = formation thickness, ft
i = injection rate, B/D
i_{base} = injection rate of fluid having same mobility as reservoir oil
i_w = water injection rate, B/D
I = injectivity index, BWPD/psi
$J(S_w)$ = dimensionless capillary pressure, defined by Leverett (see Chapter 2)
k = absolute permeability, md

$\overline{k}$ = average permeability, md
k_g = effective permeability to gas, md
k_o = effective permeability to oil, md
k_{rg} = relative permeability to gas, fraction
k_{ro} = relative permeability to oil, fraction
k_{rw} = relative permeability to water, fraction
k_w = effective permeability to water, md
k_w/k_o = water-oil permeability ratio, dimensionless
k_σ = permeability at 84.1 percent of the total sample (see Chapter 6)
ln = natural logarithm, base e
L = length, cm
M = mobility ratio, mobility of displacing fluid divided by mobility of oil, dimensionless
n = number of samples
N = initial oil in place, bbl
N_p = cumulative produced oil volume, bbl
ΔN_p = incremental produced oil volume, bbl
p = pressure, psi
p_a = pressure at abandonment, psi
p_b = bubble-point or saturation pressure, psi
p_e = pressure at external boundary, psi
p_f = front or interface pressure, psi
p_g = pressure in gas phase, atm
p_i = initial pressure, psi
p_{iwf} = injection well bottom-hole pressure, psi
p_o = pressure in oil phase, atm
p_p = producing well bottom-hole pressure, psi
$\overline{p}_R$ = average reservoir pressure, psi
p_w = bottom-hole pressure, psi, or pressure in water phase, atm
P_c = capillary pressure, equal to $(p_o - p_w)$, atm
q = flow rate, cc/sec, or production rate, B/D
q_o = oil production rate, BOPD
q_w = water production rate, BWPD
Q_i = pore volumes of cumulative injected fluid
r = radial distance, ft
r_e = external boundary radius, ft
r_w = well radius, ft
R = ratio of producing rates of corner well to that of side well, inverted nine-spot pattern, in Table 5.12
R_s = solution GOR, scf/bbl

S_d = displacing phase saturation, fraction pore volume

$\overline{S}_d$ = average displacing phase saturation, fraction pore volume

S_g = gas saturation, fraction pore volume

S_g^* = gas saturation, in pilot flooding, equal to $\frac{\pi}{4}(1 - S_{wc} - S_{or})$

S_{gi} = initial gas saturation, fraction pore volume

S_{gt} = trapped gas saturation, fraction pore volume

S_o = oil saturation, fraction pore volume

S_{oi} = initial oil saturation, fraction pore volume

S_{or} = residual oil saturation, fraction pore volume

S_w = water saturation, fraction pore volume

$\overline{S}_w$ = average water saturation, fraction pore volume

S_{wc} = connate water saturation, fraction pore volume

S_{wi} = initial water saturation, fraction pore volume

t = time, days

u = Darcy velocity, flow rate per unit gross area, cm/sec; i.e. q/A

V = permeability variation, dimensionless

V_d = displaceable volume, equal to the cumulative injected fluid as a fraction of the product of the pattern pore volume and the displacement efficiency of the flood

V_P = pore volume, bbl

W = water volume, bbl

W_i = cumulative injected water volume, bbl

ΔW_i = incremental injected water volume, bbl

ΔW_p = incremental produced water volume, bbl

x = horizontal dimension

y = vertical dimension

α_d = angle of formation dip, degrees

θ_c = contact angle, degrees

γ = conductance ratio, ratio of injectivity of water at any pattern areal sweep to that of reservoir oil

λ = mobility (k/μ)

λ_g = gas mobility (k_g/μ_g)

λ_o = oil mobility (k_o/μ_o)

λ_w = water mobility (k_w/μ_w)

μ = viscosity, cp

μ_g = gas viscosity, cp

μ_o = oil viscosity, cp

μ_w = water viscosity, cp

ρ = density, gm/cc

ρ_g = gas density, gm/cc

ρ_o = oil density, gm/cc

ρ_w = water density, gm/cc

$\Delta\rho$ = density difference, water density minus oil density, gm/cc

σ = surface or interfacial tension, dynes/cm; also standard deviation (in statistics)

ϕ = porosity, fraction

ψ_s = fraction of total flow coming from the swept portion of the pattern

Subscripts

a = atmospheric

bt = breakthrough

c = corner well

cf = crossflow

d = displacing phase

f = front or fillup

fo = floodout

g = gas

i = initial value, injection, or interference

i,c = injection well—corner producing well, nine-spot pattern

i,s = injection well—side producing well, nine-spot pattern

ncf = no cross flow

o = oil

os = oil-solid

ow = oil-water

p = produced

r = relative or residual

s = side well or previously swept

sz = stabilized zone

t = total

u = uniform or not previously swept (i.e., unswept)

w = water

ws = water-solid

x = horizontal direction

y = vertical direction

2 = producing end

Bibliography

Abernathy, B. F.: "Waterflood Prediction Methods Compared to Pilot Performance in Carbonate Reservoirs", *J. Pet Tech.* (March, 1964) 276-282.

Ache, P. S.: "Inclusion of Radial Flow in Use of Permeability Distribution in Waterflood Calculations", Paper 935-G presented at SPE 32nd Annual Fall Meeting, Dallas, Tex., Oct. 6-9, 1957.

Adams, N. K.: *The Physics and Chemistry of Surfaces*, Oxford U. Press, London (1941).

Agan, J. B. and Fernandes, R. J.: "Performance Prediction of a Miscible Slug Process in a Highly Stratified Reservoir", *J. Pet. Tech.* (Jan., 1962) 81-86.

Alpay, O.: "A Study of Physical and Textural Heterogeneity in Sedimentary Rocks", PhD dissertation, Purdue U., Lafayette, Ind. (June, 1963).

Amott, E.: "Observations Relating to the Wettability of Porous Rock", *Trans.*, AIME (1959) **216**, 156-162.

Arman, I. H.: "Relative Permeability Studies", MS thesis, The U. of Oklahoma, Norman (1952).

Arnold, M. D., Gonzalez, H. J. and Crawford, P. B.: "Estimation of Reservoir Anisotropy from Production Data", *J. Pet. Tech.* (Aug., 1962) 909-912.

Aronofsky, J.: "Mobility Ratio, Its Influence on Flood Patterns During Water Encroachment", *Trans.*, AIME (1952) **195**, 15-24.

Aronofsky, J. S. and Ramey, H. J., Jr.: "Mobility Ratio — Its Influence on Injection and Production Histories in Five-Spot Water Flood", *Trans.*, AIME (1956) **207**, 205-210.

Arps, J. J.: "Estimation of Primary Oil Reserves", *Trans.*, AIME (1956) **207**, 182-191.

Arps, J. J., Brons, F., van Everdingen, A. F., Buchwald, R. W. and Smith, A. E.: "A Statistical Study of Recovery Efficiency", *Bull. 14D*, API (1967).

Bail, P. T.: "The Calculation of Water Flood Performance for the Bradford Third Sand from Relative Permeability and Capillary Pressure Data", *Prod. Monthly* (July, 1956) **21**, No. 7, 20-27.

Beeson, D. M. and Ortloff, G. D.: "Laboratory Investigation of the Water-Driven Carbon Dioxide Process for Oil Recovery", *Trans.*, AIME (1959) **216**, 388-391.

Benner, F. C. and Bartell, F. E.: "The Effect of Polar Impurities Upon Capillary and Surface Phenomena in Petroleum Production", *Drill. and Prod. Prac.*, API (1941).

Benner, F. C., Riches, W. W. and Bartell, F. E.: "Nature and Importance of Surface Forces in Production of Petroleum", *Drill. and Prod. Prac.*, API (1939).

Bennion, D. W. and Griffiths, J. C.: "A Stochastic Model for Predicting Variations in Reservoir Rock Properties", *Soc. Pet. Eng. J.* (March, 1966) 9-16.

Bernard, W. J. and Caudle, B. H.: "Model Studies of Pilot Waterfloods", *J. Pet. Tech.* (March, 1967) 404-410.

Bobek, J. E., Mattax, C. C. and Denekas, M. O.: "Reservoir Rock Wettability — Its Significance and Evaluation", *Trans.*, AIME (1958) **213**, 155-160.

Botset, H. G.: "The Electrolytic Model and Its Application to the Study of Recovery Problems", *Trans.*, AIME (1946) **165**, 15-25.

Bradley, H. B., Heller, J. P. and Odeh, A. S.: "A Potentiometric Study of the Effects of Mobility Ratio on Reservoir Flow Patterns", *Soc. Pet. Eng. J.* (Sept., 1961) 125-129.

Brown, H. W.: "Capillary Pressure Investigations", *Trans.*, AIME (1951) **192**, 67-74.

Brown, R. J. S. and Fatt, I.: "Measurements of Fractional Wettability of Oilfield Rocks by the Nuclear Magnetic Relaxation Method", *Trans.*, AIME (1956) **207**, 262-264.

Brown, W. O.: "The Mobility of Connate Water During a Water Flood", *Trans.*, AIME (1957) **210**, 190-195.

Buckley, S. E. and Leverett, M. C.: "Mechanism of Fluid Displacement in Sands", *Trans.*, AIME (1942) **146**, 107-116.

Buckwalter, J. F., Edgerton, G. H., Stiles, W. E., Earlougher, R. C., Buckles, G. L. and Bridges, P. M.: "Waterflood Oil Recovery Is Lessened by Restricting Rates", *Oil and Gas J.* (June 15, 1958) 88-89.

Burdine, N. T.: "Relative Permeability Calculations from Pore Size Distribution Data", *Trans.*, AIME (1953) **198**, 71-78.

Burton, M. B., Jr., and Crawford, P. B.: "Application of the Gelatin Model for Studying Mobility Ratio Effects", *Trans.*, AIME (1956) **207**, 333-337.

Bush, J. L. and Helander, D. P.: "Empirical Prediction of Recovery Rate in Waterflooding Depleted Sands", *J. Pet. Tech.* (Sept., 1968) 933-943.

Calhoun, J. C., Jr.: *Fundamentals of Reservoir Engineering*, The U. of Oklahoma Press, Norman (1960).

Calloway, F. H.: "Evaluation of Waterflood Prospects", *J. Pet. Tech.* (Oct., 1959) 11-15.

Cardwell, W. T., Jr.: "The Meaning of the Triple Value in Non-Capillary Buckley-Leverett Theory", *Trans.*, AIME (1959) **216**, 271-276.

Carll, John F.: *The Geology of the Oil Regions of Warren, Venango, Clarion, and Butler Counties, Pennsylvania*, Second Geological Survey of Pennsylvania (1880) **III**, 1875-1879.

Carpenter, C. W., Jr., Bail, P. T. and Bobek, J. E.: "A Verification of Waterflood Scaling in Heterogeneous Communicating Flow Models", *Soc. Pet. Eng. J.* (March, 1962) 9-12.

Caudle, B. H.: "Injection Pattern Sweepout Efficiency", *Pet. Eng.* (June, 1959) **31**, B34.

Caudle, B. H. and Dyes, A. B.: "Improving Miscible Displacement by Gas-Water Injection", *Trans.*, AIME (1958) **213**, 281-284.

Caudle, B. H., Erickson, R. A. and Slobod, R. L.: "The Encroachment of Injected Fluids Beyond the Normal Well Pattern", *Trans.*, AIME (1955) **204**, 79-85.

Caudle, B. H., Hickman, B. M. and Silberberg, I. H.: "Performance of the Skewed Four-Spot Injection Pattern", *J. Pet. Tech.* (Nov., 1968) 1315-1319.

Caudle, B. H. and Loncaric, I. G.: "Oil Recovery in Five-Spot Pilot Floods", *Trans.*, AIME (1960) **219**, 132-136.

Caudle, B. H., Slobod, R. L. and Brownscombe, E. R.: "Further Developments in the Laboratory Determination of Relative Permeability", *Trans.*, AIME (1951) **192**, 145-150.

Caudle, B. H. and Witte, M. D.: "Production Potential Changes During Sweep-Out in a Five-Spot System", *Trans.*, AIME (1959) **216**, 446-448.

Chatenever, A. and Calhoun, J. C., Jr.: "Visual Examinations of Fluid Behavior in Porous Media — Part I", *Trans.*, AIME (1952) **195**, 149-156.

Cheek, R. E. and Menzie, D. E.: "Fluid Mapper Model Studies of Mobility Ratio", *Trans.*, AIME (1955) **204**, 278-281.

Chuoke, R. L., van Meurs, P. and van der Poel, C.: "The Instability of Slow Immiscible Viscous Liquid-Liquid Displacements in Permeable Media", *Trans.*, AIME (1959) **216**, 188-194.

Clark, K. K.: "Pressure Transient Testing of Fractured Water Injection Wells", *J. Pet. Tech.* (June, 1968) 639-643.

Corey, A. T.: "The Interrelation Between Gas and Oil Relative Permeabilities", *Prod. Monthly* (Nov., 1954) **19**, No. 11, 34-41.

Corey, A. T., Rathjens, C. H., Henderson, J. H. and Wyllie, M. R. J.: "Three Phase Relative Permeability", *Trans.*, AIME (1956) **207**, 349-351.

Cotman, N. T., Still, G. R. and Crawford, P. B.: "Laboratory Comparison of Oil Recovery in Five-Spot and Nine-Spot Waterflood Patterns", *Prod. Monthly* (Dec., 1962) **27**, No. 12, 10-13.

Craig, F. F., Jr.: Discussion of "Combination Method for Predicting Waterflood Performance for Five-Spot Patterns in Stratified Reservoirs", *J. Pet. Tech.* (Feb., 1969) 233-234.

Craig, F. F., Jr.: "Effect of Permeability Variation and Mobility Ratio on Five-Spot Oil Recovery Performance Calculations", *J. Pet. Tech.* (Oct., 1970) 1239-1245.

Craig, F. F., Jr.: "Laboratory Model Study of Single Five-Spot and Single Injection Well Pilot Waterflooding", *J. Pet. Tech.* (Dec., 1965) 1454-1460.

Craig, F. F., Jr., Geffen, T. M. and Morse, R. A.: "Oil Recovery Performance of Pattern Gas or Water Injection Operations from Model Tests", *Trans.*, AIME (1955) **204**, 7-15.

Craig, F. F., Jr., Sanderlin, J. L., Moore, D. W. and Geffen, T. M.: "A Laboratory Study of Gravity Segregation in Frontal Drives", *Trans.*, AIME (1957) **210**, 275-282.

Crawford, P. B.: "Laboratory Factors Affecting Water Flood Pattern Performance and Selection", *J. Pet. Tech.* (Dec., 1960) 11-15.

Crawford, P. B. and Collins, R. E.: "Analysis of Flooding Horizontally Fractured Thin Reservoirs", *World Oil* (1954) Aug., 139; Sept., 173; Oct., 214; Nov., 212; and Dec., 197.

Crawford, P. B. and Collins, R. E.: "Estimated Effect of Vertical Fractures on Secondary Recovery", *Trans.*, AIME (1954) **201**, 192-196.

Crawford, P. B., Pinson, J. M., Simmons, J. and Landrum, B. L.: "Sweep Efficiencies of Vertically Fractured Five-Spot Patterns", *Pet. Eng.* (March, 1956) **28**, B95-102.

Croes, G. A. and Schwarz, N.: "Dimensionally Scaled Experiments and the Theories on the Water-Drive Process", *Trans.*, AIME (1955) **204**, 35-42.

Crowell, D. C., Dean, G. W. and Loomis, A. G.: "Efficiency of Gas Displacement from a Water Drive Reservoir", RI 6735, USBM (1966).

Dalton, R. L., Jr., Rapoport, L. A. and Carpenter, C. W., Jr.: "Laboratory Studies of Pilot Water Floods", *Trans.*, AIME (1960) **219**, 24-30.

Denekas, M. O., Mattax, C. C. and Davis, G. T.: "Effects of Crude Oil Components on Rock Wettability", *Trans.*, AIME (1959) **216**, 330-333.

Deppe, J. C.: "Injection Rates — The Effect of Mobility Ratio, Area Swept, and Pattern", *Soc. Pet. Eng. J.* (June, 1961) 81-91.

Devlikamov, V. V., Sukhanov, G. N. and Bul'chuk, D. D.: "The Nature of Oil Displacement by Water Along the Thickness of a Formation", *Dzvest. Vysshikh. Ucheb Zaved.*, Neft i Gaz (1960) **3**, No. 1, 53-57.

Dietz, D. N.: "A Theoretical Approach to the Problem of Encroaching and Bypassing Edge Water", *Proc.*, Koninkl. Ned. Akad. Wetenschap (1953) B56, 38.

Doepel, G. W. and Sibley, W. P.: "Miscible Displacement — A Multilayer Technique for Predicting Reservoir Performance", *J. Pet. Tech.* (Jan., 1962) 73-80.

Donohue, D. A. T.: "Can Aerial Photography Improve Oil Recovery?" *Oil and Gas J.* (July 10, 1967) 201-204.

Dougherty, E. L. and Sheldon, J. W.: "The Use of Fluid-Fluid Interfaces to Predict the Behavior of Oil Recovery Processes", *Soc. Pet. Eng. J.* (June, 1964) 171-182.

Douglas, J., Jr., Blair, P. M., and Wagner, R. J.: "Calculation of Linear Waterflood Behavior Including the Effects of Capillary Pressure", *Trans.*, AIME (1958) **213**, 96-102.

Douglas, J., Jr., Peaceman, D. W. and Rachford, H. H., Jr.: "A Method for Calculating Multi-Dimensional Immiscible Displacement", *Trans.*, AIME (1959) **216**, 297-306.

Dyes, A. B.: Discussion of "Mobility Ratio, Its Influence on Flood Patterns During Water Encroachment", *Trans.*, AIME (1952) **195**, 22-23.

Dyes, A. B.: "Production of Water-Driven Reservoirs Below Their Bubble Point", *Trans.*, AIME (1954) **201**, 240-244.

Dyes, A. B., Caudle, B. H. and Erickson, R. A.: "Oil Production After Breakthrough — As Influenced by Mobility Ratio", *Trans.*, AIME (1954) **201**, 81-86.

Dyes, A. B., Kemp, C. E. and Caudle, B. H.: "Effect of Fractures on Sweep-Out Pattern", *Trans.*, AIME (1958) **213**, 245-249.

Dykstra, H. and Parsons, R. L.: "The Prediction of Oil Recovery by Waterflooding", *Secondary Recovery of Oil in the United States*, 2nd ed., API (1950) 160-174.

Dykstra, H. and Parsons, R. L.: "The Prediction of Waterflood Performance with Variation in Permeability Profile", *Prod. Monthly* (1950) **15**, 9-12.

Efros, D. A.: "The Displacement of a Two-Component Mixture When the Viscosity of One of the Fluids Being Displaced Is Low", *Dokladi Akad. Nauk SSR* (July-Aug., 1958) **121**, 59-62.

Elkins, L. F.: Unpublished communication (1943) cited by S. J. Pirson in *Elements of Oil Reservoir Engineering*, McGraw-Hill Book Co., Inc., New York (1950) 328.

Elkins, L. F.: "Fosterton Field — An Unusual Problem of Bottom Water Coning and Volumetric Water Invasion Efficiency", *Trans.*, AIME (1959) **216**, 130-139.

Elkins, L. F. and Skov, A. M.: "Cyclic Waterflooding the Spraberry Utilizes 'End Effects' to Increase Oil Production Rate", *J. Pet. Tech.* (Aug., 1963) 877-883.

Elkins, L. F. and Skov, A. M.: "Determination of Fracture Orientation from Pressure Interference", *Trans.*, AIME (1960) **219**, 301-304.

Elkins, L. F. and Skov, A. M.: "Some Field Observations of Heterogeneity of Reservoir Rocks and Its Effects on Oil Displacement Efficiency", paper SPE-282 presented at SPE Production Research Symposium, Tulsa, Okla., April 12-13, 1962.

Engelberts, W. L. and Klinkenberg, L. J.: "Laboratory Experiments on the Displacement of Oil by Water from Packs of Granular Materials", *Proc.*, Third World Pet. Cong. (1951) **II**, 544.

Fatt, I.: "The Effect of Overburden Pressure on Relative Permeability", *Trans.*, AIME (1953) **198**, 325-326.

Fatt, I.: "The Network Model of Porous Media", *Trans.*, AIME (1956) **207**, 144-181.

Fatt, I. and Dykstra, H.: "Relative Permeability Studies", *Trans.*, AIME (1951) **192**, 249-256.

Fay, C. H. and Prats, M.: "The Application of Numerical Methods to Cycling and Flooding Problems", *Proc.*, Third World Pet. Cong. (1951) **II**, 555-563.

Fayers, F. J. and Sheldon, J. W.: "The Effect of Capillary Pressure and Gravity on Two-Phase Fluid Flow in a Porous Medium", *Trans.*, AIME (1959) **216**, 147-155.

Felsenthal, M., Cobb, T. R. and Heuer, G. J.: "A Comparison of Waterflood Evaluation Methods", paper 332-G presented at SPE 37th Annual Fall Meeting, Los Angeles, Calif., Oct. 7-10, 1962.

Felsenthal, M. and Yuster, S. T.: "A Study of the Effect of Viscosity in Oil Recovery by Waterflooding", paper 163-G presented at SPE West Coast Meeting, Los Angeles, Oct. 25-26, 1951.

Ferrell, H., Irby, T. L., Pruitt, G. T. and Crawford, P. B.: "Model Studies for Injection-Production Well Conversion During a Line Drive Water Flood", *Trans.*, AIME (1960) **219**, 94-98.

Fettke, C. R.: "Bradford Oil Field, Pennsylvania and New York", Pennsylvania Geological Survey, 4th Series (1938) M-21.

"Final Report — Water Flood Study, Redwater Pool", presented to Oil and Gas Conservation Board of Alberta by Canadian Gulf Oil Co., May 22, 1953.

Fitch, R. A. and Griffith, J. D.: "Experimental and Calculated Performance of Miscible Floods in Stratified Reservoirs", *J. Pet. Tech.* (Nov., 1964) 1289-1298.

"Fluid Distribution in Porous Systems — A Preview of the Motion Picture", Stanolind Oil and Gas Co. (1952); subsequently reprinted by Pan American Petroleum Corp. and Amoco Production Co.

Fried, A. N.: "The Foam-Drive Process for Increasing the Recovery of Oil", RI 5866, USBM (1961).

Funk, E. E.: "Effects of Production Restrictions on Water-Flood Recovery", *Prod. Monthly* (May, 1956) **21**, No. 5, 20-27.

Gatlin, C. and Slobod, R. L.: "The Alcohol Slug Process for Increasing Oil Recovery", *Trans.*, AIME (1960) **219**, 46-53.

Gaucher, D. H. and Lindley, D. C.: "Waterflood Performance in a Stratified Five-Spot Reservoir — A Scaled-Model Study", *Trans.*, AIME (1960) **219**, 208-215.

Geertsma, J., Croes, G. A. and Schwarz, N.: "Theory of Dimensionally Scaled Models of Petroleum Reservoirs", *Trans.*, AIME (1956) **207**, 118-127.

Geffen, T. M., Owens, W. W., Parrish, D. R. and Morse, R. A.: "Experimental Investigation of Factors Affecting Laboratory Relative Permeability Measurements", *Trans.*, AIME (1951) **192**, 99-110.

Geffen, T. M., Parrish, D. R., Haynes, G. W. and Morse, R. A.: "Efficiency of Gas Displacement from Porous Media by Liquid Flooding", *Trans.*, AIME (1952) **195**, 29-38.

Goddin, C. S., Jr., Craig, F. F., Jr., Wilkes, J. O. and Tek, M. R.: "A Numerical Study of Waterflood Performance in a Stratified System with Crossflow", *J. Pet. Tech.* (June, 1966) 765-771.

Gogarty, W. B. and Tosch, W. C.: "Miscible-Type Waterflooding: Oil Recovery with Micellar Solutions", *J. Pet. Tech.* (Dec., 1968) 1407-1414.

Gottfried, B. S., Guilinger, W. H. and Snyder, R. W.: "Numerical Solutions for the Equations for One-Dimensional Multi-Phase Flow in Porous Media", *Soc. Pet. Eng. J.* (March, 1966) 62-72.

Greenkorn, R. A., Johnson, C. R. and Stallenberger, L. K.: "Directional Permeability of Heterogeneous Anisotropic Porous Media", *Soc. Pet. Eng. J.* (June, 1964) 124-132.

Groult, J., Reiss, L. H. and Montadert, L.: "Reservoir Inhomogeneities Deduced from Outcrop Observations and Production Logging", *J. Pet. Tech.* (July, 1966) 883-891.

Guckert, L. G.: "Areal Sweepout Performance of Seven- and Nine-Spot Flood Patterns", MS thesis, The Pennsylvania State U., University Park (Jan., 1961).

Guerrero, E. T. and Earlougher, R. C.: "Analysis and Comparison of Five Methods Used to Predict Waterflooding Reserves and Performance", *Drill. and Prod. Prac.*, API (1961) 78-95.

Gurses, B. and Helander, D. P.: "Shape Factor Analysis for Peripheral Waterflood Prediction by the Channel Flow Technique", *Prod. Monthly* (April, 1967) **32**, No. 4, 2-31.

Guthrie, R. K. and Greenberger, M. H.: "The Use of Multiple-Correlation Analyses for Interpreting Petroleum Engineering Data", *Drill. and Prod. Prac.*, API (1955) 130-137.

Habermann, B.: "The Efficiency of Miscible Displacement As a Function of Mobility Ratio", *Trans.*, AIME (1960) **219**, 264-272.

Hartsock, J. H. and Slobod, R. L.: "The Effect of Mobility Ratio and Vertical Fractures on the Sweep Efficiency of a Five-Spot", *Prod. Monthly* (Sept., 1961) **26**, No. 9, 2-7.

Hassler, G. L.: U. S. Patent No. 2,345,935 (April, 1944).

Hassler, G. L., Rice, R. R. and Leeman, E. H.: "Investigations on the Recovery of Oil from Sandstones by Gas Drive", *Trans.*, AIME (1936) **118**, 116-137.

Hauber, W. C.: "Prediction of Waterflood Performance for Arbitrary Well Patterns and Mobility Ratios", *J. Pet. Tech.* (Jan., 1964) 95-103.

Hele-Shaw, H. S.: "Experiments on the Nature of the Surface Resistance in Pipes and on Ships", *Trans.*, Institution of Naval Architects (1897) **XXXIX**, 145.

Hendrickson, G. E.: "History of the Welch Field San Andres Pilot Waterflood", *J. Pet. Tech.* (Aug., 1961) 745-749.

Henley, D. H.: "Method for Studying Waterflooding Using Analog, Digital, and Rock Models", paper presented at 24th Technical Conference on Petroleum, The Pennsylvania State U., University Park (Oct., 1963).

Hiatt, W. N.: "Injected-Fluid Coverage of Multi-Well Reservoirs With Permeability Stratification", *Drill. and Prod. Prac.*, API (1958) 165-194.

Hickok, C. W., Christensen, R. J. and Ramsay, H. J., Jr.: "Progress Review of the K&S Carbonated Waterflood Project", *J. Pet. Tech.* (Dec., 1960) 20-24.

Higgins, R. V., Boley, D. W. and Leighton, A. J.: "Aids in Forecasting the Performance of Water Floods", *J. Pet. Tech.* (Sept., 1964) 1076-1082.

Higgins, R. V. and Leighton, A. J.: "A Computer Method to Calculate Two-Phase Flow in Any Irregularly Bounded Porous Medium", *J. Pet. Tech.* (June, 1962) 679-683.

Higgins, R. V. and Leighton, A. J.: "Computer Prediction of Water Drive of Oil and Gas Mixtures Through Irregularly Bounded Porous Media — Three-Phase Flow", *J. Pet. Tech.* (Sept., 1962) 1048-1054.

Higgins, R. V. and Leighton, A. J.: "Computer Techniques for Predicting Three-Phase Flow in Five-Spot Waterfloods", RI 7011, USBM (Aug., 1967).

Higgins, R. V. and Leighton, A. J.: "Waterflood Prediction of Partially Depleted Reservoirs", paper SPE 757 presented at SPE 33rd Annual California Regional Fall Meeting, Santa Barbara, Oct. 24-25, 1963.

History of Petroleum Engineering, API, Dallas, Tex. (1961).

Holbrook, O. C. and Bernard, G. G.: "Determination of Wettability by Dye Adsorption", *Trans.*, AIME (1958) **213**, 261-264.

Holmgren, C. R.: "Some Results of Gas and Water Drives on a Long Core", *Trans.*, AIME (1948) **179**, 103-118.

Holmgren, C. R. and Morse, R. A.: "Effect of Free Gas Saturation on Oil Recovery by Waterflooding", *Trans.*, AIME (1951) **192**, 135-140.

Hough, E. W., Rzasa, M. J. and Wood, B. B., Jr.: "Interfacial Tensions at Reservoir Pressures and Temperatures, Apparatus and the Water-Methane System", *Trans.*, AIME (1950) **192**, 57-62.

Hovanessian, S. A. and Fayers, F. J.: "Linear Water Flood with Gravity and Capillary Effects", *Soc. Pet. Eng. J.* (March, 1961) 32-36.

"How Temperature Affects Viscosity of Salt Water", *World Oil* (Aug. 1, 1967) 68.

Hunter, Z. Z.: "8½ Million Extra Barrels in 6 Years", *Oil and Gas J.* (Aug. 27, 1956) 92.

Hunter, Z. Z.: "Progress Report, North Burbank Unit Water Flood — Jan. 1, 1956", *Drill. and Prod. Prac.*, API (1956) 262.

Hurst, W.: "Determination of Performance Curves in Five-Spot Waterflood", *Pet. Eng.* (1953) **25**, B40-46.

Hutchinson, C. A., Jr.: "Reservoir Inhomogeneity Assessment and Control", *Pet. Eng.* (Sept., 1959) **31**, B19-26.

Hutchinson, C. A., Jr., Polasek, T. L., Jr., and Dodge, C. F.: "Identification, Classification and Prediction of Reservoir Nonuniformities Affecting Production Operations", *J. Pet. Tech.* (March, 1961) 223-230.

Jacquard, P. and Jain, C.: "Permeability Distribution From Field Pressure Data", *Soc. Pet. Eng. J.* (Dec., 1965) 281-294.

Jahns, J. O.: "A Rapid Method for Obtaining a Two-Dimensional Reservoir Description from Well Pressure Response Data", *Soc. Pet. Eng. J.* (Dec., 1966) 315-327.

Jennings, H. Y., Jr.: "Waterflood Behavior of High Viscosity Crudes in Preserved Soft and Unconsolidated Cores", *J. Pet. Tech.* (Jan., 1966) 116-120.

Johansen, R. T. and Dunning, H. N.: "Relative Wetting Tendencies of Crude Oil by the Capillarimetric Method", *Prod. Monthly* (Sept., 1959) **24**, No. 9.

Johnson, C. E., Jr.: "Prediction of Oil Recovery by Water Flood — A Simplified Graphical Treatment of the Dykstra-Parsons Method", *Trans.*, AIME (1956) **207**, 345-346.

Johnson, C. R., Greenkorn, R. A. and Woods, E. G.: "Pulse Testing: A New Method for Describing Reservoir Flow Properties Between Wells", *J. Pet. Tech.* (Dec., 1966) 1599-1604.

Johnson, E. F., Bossler, D. P. and Naumann, V. O.: "Calculation of Relative Permeability from Displacement Experiments", *Trans.*, AIME (1959) **216**, 370-372.

Johnson, J. P.: "Predicting Waterflood Performance by the Graphical Representation of Porosity and Permeability Distribution", *J. Pet. Tech.* (Nov., 1965) 1285-1290.

Johnston, N. and van Wingen, N.: "Recent Laboratory Investigation of Water Flooding in California", *Trans.*, AIME, (1953) **198**, 219-224.

Jones-Parra, J.: "Comments on Capillary Equilibrium", *Trans.*, AIME (1953) **198**, 314-316.

Jones-Parra, J. and Calhoun, J. C., Jr.: "Computation of a Linear Flood by the Stabilized Zone Method", *Trans.*, AIME (1953) **189**, 335-338.

Jordan, J. K., McCardell, W. M. and Hocott, C. R.: "Effect of Rate on Oil Recovery by Water Flooding", *Oil and Gas J.* (May 13, 1957) 99-108.

Kantar, K. and Helander, D. P.: "Graphical Technique for Interpolating Shape Factors for Peripheral Waterflood Systems", *Prod. Monthly* (June, 1967) **32**, No. 6, 2-5.

Kantar, K. and Helander, D. P.: "Simplified Shape Factor Analysis as Applied to an Unconfined Peripheral Waterflooding System", *Prod. Monthly* (May, 1967) **32**, No. 5, 14-17.

Kelley, D. L. and Caudle, B. H.: "The Effect of Connate Water on the Efficiency of High Viscosity Waterfloods", *J. Pet. Tech.* (Nov., 1966) 1481-1486.

Kennedy, H. T. and Guerrero, E. T.: "The Effect of Surface and Interfacial Tensions on the Recovery of Oil by Water Flooding", *Trans.*, AIME (1954) **201**, 124-131.

Kereluk, M. J. and Crawford, P. B.: "Comparison Between Observed and Calculated Sweeps in Oil Displacement from Stratified Reservoirs", paper presented at API Mid-Continent District Meeting, Amarillo, Tex., April 3-5, 1968.

Kern, L. R.: "Displacement Mechanism in Multi-Well Systems", *Trans.*, AIME (1952) **195**, 39-46.

Killins, C. R., Nielsen, R. F. and Calhoun, J. C., Jr.: "Capillary Desaturation and Imbibition in Rocks", *Prod. Monthly* (Feb., 1953) **18**, No. 2, 30-39.

Kimbler, O. K., Caudle, B. H. and Cooper, H. E., Jr.: "Areal Sweepout Behavior in a Nine-Spot Injection Pattern", *J. Pet. Tech.* (Feb., 1964) 199-202.

Koeller, R. C. and Craig, F. F., Jr.: Discussion of "Mobility Ratio — Its Influence on Injection and Production Histories in Five-Spot Water Flood", *Trans.*, AIME (1956) **207**, 291-292.

Kruger, W. D.: "Determining Areal Permeability Distribution by Calculations", *J. Pet. Tech.* (July, 1961) 691-696.

Krutter, H.: "Nine-Spot Flooding Program", *Oil and Gas J.* (Aug. 17, 1939) **38**, No. 14, 50.

Kufus, H. B. and Lynch, E. J.: "Linear Frontal Displacement in Multilayer Sands", *Prod. Monthly* (Dec., 1959) **24**, No. 12, 32-35.

Kyte, J. R., Naumann, V. O. and Mattax, C. C.: "Effect of Reservoir Environment on Water-Oil Displacements", *J. Pet. Tech.* (June, 1961) 579-582.

Kyte, J. R. and Rapoport, L. A.: "Linear Waterflood Behavior and End Effects in Water-Wet Porous Media", *Trans.*, AIME (1958) **213**, 423-426.

Kyte, J. R., Stanclift, R. J., Jr., Stephan, S. C., Jr., and Rapoport, L. A.: "Mechanism of Water Flooding in the Presence of Free Gas", *Trans.*, AIME (1956) **207**, 215-221.

"Laboratory Investigation of the Effect of Free Gas in Waterflooding Redwater Cores", report presented to Oil and Gas Conservation Board of Alberta by Imperial Oil Ltd., Oct., 1955.

"Laboratory Investigation of the Effect of Free Gas in Waterflooding Redwater Cores", report presented to Oil and Gas Conservation Board of Alberta by Imperial Oil Ltd., March, 1956.

Land, C. S.: "Calculation of Imbibition Relative Permeability for Two- and Three-Phase Flow from Rock Properties", *Soc. Pet. Eng J.* (June, 1968) 149-156.

Landrum, B. L. and Crawford, P. B.: "Effect of Directional Permeability on Sweep Efficiency and Production Capacity", *Trans.*, AIME (1960) **219**, 407-411.

Landrum, B. L. and Crawford, P. B.: "Estimated Effect of Horizontal Fractures in Thick Reservoirs on Pattern Conductivity", *Trans.*, AIME (1957) **210**, 399-401.

Law, J.: "Statistical Approach to the Interstitial Heterogeneity of Sand Reservoirs", *Trans.*, AIME (1944) **155**, 202-222.

Leach, R. O.: "Surface Equilibrium in Contact Angle Measurements", paper presented at Gordon Research Conference on Chemistry at Interfaces, Meridan, N.H. (July, 1957).

Leach, R. O., Wagner, O. R., Wood, H. W. and Harpke, C. F.: "A Laboratory and Field Study of Wettability Adjustment in Waterflooding", *J. Pet. Tech.* (Feb., 1962) 205-212.

Leas, W. J., Jenks, L. H. and Russell, C. D.: "Relative Permeability to Gas", *Trans.*, AIME (1950) **189**, 65-72.

Lee, B. D.: "Potentiometric Model Studies of Fluid Flow in Petroleum Reservoirs", *Trans.*, AIME (1948) **174**, 41-66.

Leverett, M. C.: "Capillary Behavior in Porous Solids", *Trans.*, AIME (1941) **142**, 159-169.

Leverett, M. C. and Lewis, W. B.: "Steady Flow of Gas-Oil-Water Mixtures Through Unconsolidated Sands", *Trans.*, AIME (1941) **142**, 107-120.

Levine, J. S.: "Displacement Experiments in a Consolidated Porous System", *Trans.*, AIME (1954) **201**, 57-66.

Martin, J. C.: "Some Mathematical Aspects of Two-Phase Flow With Application to Flooding and Gravity Segregation Problems", *Prod. Monthly* (April, 1958) **23**, No. 4, 22-35.

Matthews, C. S. and Fischer, M. J.: "Effect of Dip on Five-Spot Sweep Pattern", *Trans.*, AIME (1956) **207**, 111-117.

Matthews, C. S. and Russell, D. G.: *Pressure Buildup and Flow Tests in Wells*, Monograph Series, Society of Petroleum Engineers, Dallas, Tex. (1967) **1**.

McCardell, W. M.: "Further Discussion of 'Effects of Curtailment on Waterflood Recovery'", *Pet. Eng.* (June, 1959) **31**, B128-133.

McCarty, D. G. and Barfield, E. C.: "The Use of High-Speed Computers for Predicting Flood-Out Patterns", *Trans.*, AIME (1958) **213**, 139-145.

McEwen, C. R.: "A Numerical Solution of the Linear Displacement Equation With Capillary Pressure", *Trans.*, AIME (1959) **216**, 412-415.

McLatchie, A. S., Hemstock, R. A. and Young, J. W.: "The Effective Compressibility of Reservoir Rock and Its Effect on Permeability", *Trans.*, AIME (1958) **213**, 386-388.

Menzie, D. E. and Cole, F. W.: "Waterflood Curtailment — Pro and Con", *Pet. Eng.* (May, 1960) **32**, B48-57.

Messer, E. S.: "Interstitial Water Determination by an Evaporation Method", *Trans.*, AIME (1951) **192**, 269-274.

Miller, M. G. and Lents, M. R.: "Performance of Bodcaw Reservoir, Cotton Valley Field Cycling Project: New Methods of Predicting Gas-Condensate Reservoir Performance Under Cycling Operations Compared to Field Data", *Drill. and Prod. Prac.*, API (1947) 128-149.

Moore, A. D.: "Fields From Fluid Flow Mappers", *J. Appl. Phys.* (1949) **20**, 790.

Moore, A. D.: "Mapping Technique Applied to Fluid Mapper Patterns", *Trans.*, AIEE (1952) **71**, Part I, 1-4.

Moore, A. D.: "The Further Development of Fluid Mappers", *Trans.*, AIEE (1950) **69**, Part II, 1615-1624.

Morel-Seytoux, H. J.: "Analytical-Numerical Method in Waterflooding Predictions", *Soc. Pet. Eng. J.* (Sept., 1965) 247-258.

Morel-Seytoux, H. J.: "Unit Mobility Ratio Displacement Calculations for Pattern Floods in Homogeneous Medium", *Soc. Pet. Eng. J.* (Sept., 1966) 217-227.

Morse, R. A., Terwilliger, P. L. and Yuster, S. T.: "Relative Permeability Measurements on Small Core Samples", *Oil and Gas J.* (Aug. 23, 1947); also Technical Paper 124, Pennsylvania State College Mineral Industries Experiment Station (1947).

Mortada, M. and Nabor, G. W.: "An Approximate Method for Determining Areal Sweep Efficiency and Flow Capacity in Formations With Anisotropic Permeability", *Soc. Pet. Eng. J.* (Dec., 1961) 277-286.

Moss, J. T., White, P. D. and McNiel, J. S., Jr.: "In-Situ Combustion Process — Results of a Five-Well Field Experiment", *Trans.*, AIME (1959) **216**, 55-64.

Mungan, N.: "Certain Wettability Effects in Laboratory Waterfloods", *J. Pet. Tech.* (Feb., 1966) 247-252.

Mungan, N.: "Interfacial Effects in Immiscible Liquid-Liquid Displacement in Porous Media", *Soc. Pet. Eng. J.* (Sept., 1966) 247-253.

Mungan, N.: "Role of Wettability and Interfacial Tension in Water Flooding", *Soc. Pet. Eng. J.* (June, 1964) 115-123.

Muskat, M.: *Flow of Homogeneous Fluids Through Porous Systems*, J. W. Edwards, Inc., Ann Arbor, Mich. (1946).

Muskat, M.: *Physical Principles of Oil Production*, McGraw-Hill Book Co., Inc., New York (1949).

Muskat, M.: "The Effect of Permeability Stratifications in Complete Water Drive Systems", *Trans.*, AIME (1950) **189**, 349-358.

Muskat, M.: *The Flow of Homogeneous Fluids Through Porous Media*, McGraw-Hill Book Co., Inc., New York (1937).

Muskat, M.: "The Theory of Nine-Spot Flooding Networks", *Prod. Monthly* (March, 1948) **13**, No. 3, 14.

Muskat, M. and Wyckoff, R. D.: "A Theoretical Analysis of Waterflooding Networks", *Trans.*, AIME (1934) **107**, 62-76.

Naar, J. and Wygal, R. J.: "Three-Phase Imbibition Relative Permeability", *Soc. Pet. Eng. J.* (Dec., 1961) 254-263.

Naar, J., Wygal, R. J. and Henderson, J. H.: "Imbibition Relative Permeability in Unconsolidated Porous Media", *Soc. Pet. Eng. J.* (March, 1962) 13-23.

Neilson, I. D. R. and Flock, D. L.: "The Effect of a Free Gas Saturation on the Sweep Efficiency of an Isolated Five-Spot", *Bull.*, CIM (1962) **55**, 124-129.

Newcombe, J., McGhee, J. and Rzasa, M. J.: "Wettability Versus Displacement in Water Flooding in Unconsolidated Sand Columns", *Trans.*, AIME (1955) **204**, 227-232.

Nobles, M. A. and Janzen, H. B.: "Application of a Resistance Network for Studying Mobility Ratio Effects", *Trans.*, AIME (1958) **213**, 356-358.

Nutting, P. G.: "Some Physical and Chemical Properties of Reservoir Rocks Bearing on the Accumulation and Discharge of Oil", *Problems in Petroleum Geology*, AAPG (1934).

Osoba, J. S., Richardson, J. G., Kerver, J. K., Hafford, J. A. and Blair, P. M.: "Laboratory Measurements of Relative Permeability", *Trans.*, AIME (1951) **192**, 47-56.

Overbey, W. K., Jr. and Rough, R. L.: "Surface Joint Patterns Predict Well Bore Fracture Orientation", *Oil and Gas J.* (Feb. 26, 1968) 84-86.

Owens, W. W.: Private Communication.

Owens, W. W. and Archer, D. L.: "Waterflood Pressure Pulsing for Fractured Reservoirs", *J. Pet. Tech.* (June, 1966) 745-752.

Owens, W. W., Parrish, D. R. and Lamoreaux, W. E.: "An Evaluation of a Gas Drive Method for Determining Relative Permeability Relationships", *Trans.*, AIME (1956) **207**, 275-280.

Parrish, D. R. and Craig, F. F., Jr.: "Laboratory Study of a Combination of Forward Combustion and Waterflooding — The COFCAW Process", *J. Pet. Tech.* (June, 1969) 753-761.

Patton, E. C., Jr.: "Evaluation of Pressure Maintenance by Internal Gas Injection in Volumetrically Controlled Reservoirs", *Trans.*, AIME (1947) **170**, 112-155.

Paulsell, B. L.: "Areal Sweep Performance of Five-Spot Pilot Floods", MS thesis, The Pennsylvania State U., University Park (Jan., 1958).

Perkins, F. M., Jr., and Collins, R. E.: "Scaling Laws for Laboratory Flow Models of Oil Reservoirs", *Trans.*, AIME (1960) **219**, 383-385.

Pfister, R. J.: "More Oil From Spent Water Drives by Intermittent Air or Gas Injection", *Prod. Monthly* (Sept., 1947) **12**, No. 9, 10-12.

Pickell, J. J., Swanson, B. F. and Hickman, W. B.: "Application of Air-Mercury and Oil-Air Capillary Pressure Data in the Study of Pore Structure and Fluid Distribution", *Soc. Pet. Eng. J.* (March, 1966) 55-61.

Pinson, J., Simmons, J., Landrum, B. J. and Crawford, P. B.: "Effect of Large Elliptical Fractures on Sweep Efficiencies in Water Flooding or Fluid Injection Programs", *Prod. Monthly* (Nov., 1963) **28**, No. 11, 20-22.

Pirson, S. J.: *Elements of Oil Reservoir Engineering*, McGraw-Hill Book Co., Inc., New York (1950).

Pirson, S. J., Boatman, E. M. and Nettle, R. L.: "Prediction of Relative Permeability Characteristics of Intergranular Reservoir Rocks from Electrical Resistivity Measurements", *J. Pet. Tech.* (May, 1964) 564-570.

Pottier, J. and Jacquard, P.: "Influence of Capillarity on Unstable Displacement of Immiscible Fluids in Porous Media", *Revue IFP* (April, 1963) **18**, No. 4, 527-540.

Prats, M.: "The Breakthrough Sweep Efficiency of a Staggered Line Drive", *Trans.*, AIME (1956) **207**, 361-362.

Prats, M., Hazebroek, P. and Allen, E. E.: "Effect of Off-Pattern Wells on the Behavior of a Five-Spot Flood", *Trans.*, AIME (1962) **225**, 173-178.

Prats, M., Matthews, C. S., Jewett, R. L. and Baker, J. D.: "Prediction of Injection Rate and Production History for Multifluid Five-Spot Floods", *Trans.*, AIME (1959) **216**, 98-105.

Prats, M., Strickler, W. R. and Matthews, C. S.: "Single-Fluid Five-Spot Floods in Dipping Reservoirs", *Trans.*, AIME (1955) **204**, 160-167.

Purcell, W. R.: "Capillary Pressures — Their Measurement Using Mercury and the Calculation of Permeability Therefrom", *Trans.*, AIME (1949) **186,** 39-48.

Purcell, W. R.: "Interpretation of Capillary Pressure Data", *Trans.*, AIME (1950) **189,** 369-374.

Rachford, H. H., Jr.: "Instability in Water Flooding Oil from Water-Wet Porous Media Containing Connate Water", *Soc. Pet. Eng. J.* (June, 1964) 133-148.

Ramey, H. J., Jr., and Nabor, G. W.: "A Blotter-Type Electrolytic Model Determination of Areal Sweeps in Oil Recovery by In-Situ Combustion", *Trans.*, AIME (1954) **201,** 119-123.

Rapoport, L. A.: "Scaling Laws for Use in Design and Operation of Water-Oil Flow Models", *Trans.*, AIME (1955) **204,** 143-150.

Rapoport, L. A., Carpenter, C. W., Jr., and Leas, W. J.: "Laboratory Studies of Five-Spot Waterflood Performance", *Trans.*, AIME (1958) **213,** 113-120.

Rapoport, L. A. and Leas, W. J.: "Properties of Linear Waterfloods", *Trans.*, AIME (1953) **189,** 139-148.

Raza, S. H., Treiber, L. E. and Archer, D. L.: "Wettability of Reservoir Rocks and Its Evaluation", *Prod. Monthly* (April, 1968) **33,** No. 4, 2-7.

Richardson, J. G.: "The Calculation of Waterflood Recovery from Steady State Relative Permeability Data", *Trans.*, AIME (1957) **210,** 373-375.

Richardson, J. G., Kerver, J. K., Hafford, J. A. and Osoba, J. S.: "Laboratory Determination of Relative Permeability", *Trans.*, AIME (1952) **195,** 187-196.

Richardson, J. G. and Perkins, F. M., Jr.: "A Laboratory Investigation of the Effect of Rate on Recovery of Oil by Water Flooding", *Trans.*, AIME (1957) **210,** 114-121.

Richardson, J. G., Perkins, F. M., Jr., and Osoba, J. S.: "Differences in Behavior of Fresh and Aged East Texas Woodbine Cores", *Trans.*, AIME (1955) **204,** 86-91.

Roberts, T. G.: "A Permeability Block Method of Calculating a Water Drive Recovery Factor", *Pet. Eng.* (1959) **31,** B45-48.

Root, P. J. and Skiba, F. F.: "Crossflow Effects During an Idealized Displacement Process in a Stratified Reservoir", *Soc. Pet. Eng. J.* (Sept., 1965) 229-238.

Rose, W. R. and Bruce, W. A.: "Evaluation of Capillary Character in Petroleum Reservoir Rock", *Trans.*, AIME (1949) **186,** 127-133.

Rosenbaum, M. J. F. and Matthews, C. S.: "Studies on Pilot Water Flooding", *Trans.*, AIME (1959) **216,** 316-323.

Sandberg, C. R., Gournay, L. S. and Sippel, R. F.: "The Effect of Fluid-Flow Rate and Viscosity on Laboratory Determinations of Oil-Water Relative Permeabilities", *Trans.*, AIME (1958) **213,** 36-43.

Sandiford, B. B.: "Laboratory and Field Studies of Water Floods Using Polymer Solutions to Increase Oil Recoveries", *J. Pet. Tech.* (Aug., 1964) 917-922.

Sandrea, R. J. and Farouq Ali, S. M.: "The Effects of Isolated Permeability Interferences on the Sweep Efficiency and Conductivity of a Five-Spot Network", *Soc. Pet. Eng. J.* (March, 1967) 20-30.

Saraf, D. N. and Fatt, I.: "Three Phase Relative Permeability Measurement Using a Nuclear Magnetic Resonance Technique for Estimating Fluid Saturation", *Soc. Pet. Eng. J.* (Sept., 1967) 235-242.

Sarem, A. M.: "Three Phase Relative Permeability Measurements by Unsteady-State Method", *Soc. Pet. Eng. J.* (Sept., 1966) 199-205.

Schauer, P. E.: "Application of Empirical Data in Forecasting Waterflood Behavior", paper 934-G presented at SPE 32nd Annual Fall Meeting, Dallas, Tex., Oct. 6-9, 1957.

Scheidegger, A. and Johnson, E. F.: "The Statistical Behavior of Instabilities in Displacement Processes in Porous Media", *Cdn. J. Phys.* (1961) **39,** 326-334.

Schiffman, L. and Breston, J. N.: "The Effect of Gas on Recovery When Water Flooding Long Cores", Paper No. 54, Pennsylvania State College Mineral Industries Experiment Station (Oct., 1949).

Schmalz, J. P. and Rahme, H. D.: "The Variation of Waterflood Performance with Variation in Permeability Profile", *Prod. Monthly* (Sept., 1950) **15,** No. 9, 9-12.

Schoeppel, R. J.: "Waterflood Prediction Methods", *Oil and Gas J.* (1968) Jan. 22, 72-75; Feb. 19, 98-106; March 18, 91-93; April 8, 80-86; May 6, 111-114; June 17, 100-105; July 8, 71-79.

Scott, J. O. and Forrester, C. E.: "Performance of Domes Unit Carbonate Waterflood — First Stage", *J. Pet. Tech.* (Dec., 1965) 1379-1384.

Sheffield, M.: "Three-Phase Fluid Flow Including Gravitational, Viscous, and Capillary Forces", *Soc. Pet. Eng. J.* (June, 1969) 255-269.

Simmons, J., Landrum, B. L., Pinson, J. M. and Crawford, P. B.: "Swept Areas After Breakthrough in Vertically Fractured Five-Spot Patterns", *Trans.*, AIME (1959) **216,** 73-77.

Slider, H. C.: "New Method Simplifies Predicting Waterflood Performance", *Pet. Eng.* (Feb., 1961) **33,** B68-78.

Slobod, R. L. and Blum, H. A.: "Method for Determining Wettability of Reservoir Rocks", *Trans.*, AIME (1952) **195,** 1-4.

Slobod, R. L. and Caudle, B. H.: "X-Ray Shadowgraph Studies of Areal Sweepout Efficiencies", *Trans.*, AIME (1952) **195,** 265-270.

Slobod, R. L. and Crawford, D. A.: "Evaluation of Reliability of Fluid Flow Models for Areal Sweepout Studies", *Prod. Monthly* (Oct., 1962) **27,** No. 10, 18-22.

Snell, R. W.: "Three-Phase Relative Permeability in an Unconsolidated Sand", *J. Institute of Petroleum* (March, 1962) **48,** 459.

Snyder, R. W. and Ramey, H. J., Jr.: "Application of Buckley-Leverett Displacement Theory to Noncommunicating Layered Systems", *J. Pet. Tech.* (Nov., 1967) 1500-1506.

Stewart, C. R., Craig, F. F., Jr., and Morse, R. A.: "Determination of Limestone Performance Characteristics by Model Flow Tests", *Trans.*, AIME (1953) **198,** 93-102.

Stiles, W. E.: "Use of Permeability Distribution in Water Flood Calculations", *Trans.*, AIME (1949) **186,** 9-13.

Still, G. R. and Crawford, P. B.: "Laboratory Evaluation of Oil Recovery by Cross-Flooding", *Prod. Monthly* (Feb., 1963) **28,** No. 2, 12-19.

Suder, F. E. and Calhoun, J. C., Jr.: "Waterflood Calculations", *Drill. and Prod. Prac.*, API (1949) 260-270.

Taber, J. J.: "The Injection of Detergent Slugs in Water Floods", *Trans.*, AIME (1958) **213,** 186-192.

Terwilliger, P. L., Wilsey, L. E., Hall, H. N., Bridges, P. M. and Morse, R. A.: "An Experimental and Theoretical Investigation of Gravity Drainage Performance", *Trans.*, AIME (1951) **192,** 285-296.

Testerman, D. J.: "A Statistical Reservoir Zonation Technique", *J. Pet. Tech.* (Aug., 1962) 889-893.

Torrey, P. D.: "Effects of Curtailment on Waterflood Recovery", *Pet. Eng.* (Dec., 1958) **30,** B70-74.

van Meurs, P.: "The Use of Transparent Three-Dimensional Models for Studying the Mechanism of Flow Processes in Oil Reservoirs", *Trans.*, AIME (1957) **210,** 295-301.

van Meurs, P. and van der Poel, C.: "A Theoretical Description of Water-Drive Processes Involving Viscous Fingering", *Trans.*, AIME (1958) **213,** 103-112.

Wagner, O. R. and Leach, R. O.: "Effect of Interfacial Tension on Displacement Efficiency", *Soc. Pet. Eng. J.* (Dec., 1966) 335-344.

Wagner, O. R. and Leach, R. O.: "Improving Oil Displacement Efficiency by Wettability Adjustment", *Trans.*, AIME (1959) **216,** 65-72.

Warren, J. E. and Cosgrove, J. J.: "Prediction of Waterflood Performance in a Stratified System", *Soc. Pet. Eng. J.* (June, 1964) 149-157.

Warren, J. E. and Price, H. S.: "Flow in Heterogeneous Porous Media", *Soc. Pet. Eng. J.* (Sept., 1961) 153-169.

Wasson, J. A. and Schrider, L. A.: "Combination Method for Predicting Waterflood Performance for Five-Spot Patterns in Stratified Reservoirs", *J. Pet. Tech.* (Oct., 1968) 1195-1202.

Watson, R. E., Silberberg, I. H. and Caudle, B. H.: "Model Studies of Inverted Nine-Spot Injection Pattern", *J. Pet. Tech.* (July, 1964) 801-804.

Wayhan, D. A., Albrecht, R. A., Andrea, D. W. and Lancaster, W. R.: "Estimating Waterflood Recovery in Sandstone Reservoirs", paper 875-24-A presented at Rocky Mountain District Spring Meeting, API Div. of Production, Denver, Colo., April 27-29, 1970.

Welge, H. J.: "A Simplified Method for Computing Oil Recovery by Gas or Water Drive", *Trans.*, AIME (1952) **195,** 91-98.

Welge, H. J.: "Displacement of Oil from Porous Media by Water and Gas", *Trans.*, AIME (1948) **179,** 133-138.

Wyckoff, R. D., Botset, H. G. and Muskat, M.: "The Mechanics of Porous Flow Applied to Water-Flooding Problems", *Trans.*, AIME (1933) **103,** 219-242.

Wyllie, M. R. J. and Gardner, G. H. F.: "The Generalized Kozeny-Carman Equation", *World Oil* (March and April, 1958) **146,** No. 4, 121-128 and No. 5, 210-228.

Yuster, S. T. and Calhoun, J. C., Jr.: "Behavior of Water Injection Wells", *Oil Weekly* (Dec. 18 and 25, 1944) 44-47.

Zieto, G. A.: "Interbedding of Shale Breaks and Reservoir Heterogeneities", *J. Pet. Tech.* (Oct., 1965) 1223-1228.

Author Index

(List of authors referred to in the Monograph text, references and the bibliography.)

A

Abernathy, B. F., 88, 96, 127
Ache, P. S., 85, 95, 127
Adams, N. K., 26, 127
Agan, J. B., 95, 127
Albrecht, R. A., 96, 134
Allen, E. E., 61, 132
Alpay, O., 11, 66, 67, 127
American Petroleum Institute, 11, 85, 130
Amott, E., 14, 26, 127
Andrea, D. W., 96, 134
Archer, D. L., 26, 102, 132, 133
Arman, I. H., 27, 127
Arnold, M. D., 62, 67, 127
Aronofsky, J., 45, 47, 51-53, 59, 60, 80, 85, 95, 127
Arps, J. J., 83, 85, 95, 96, 127

B

Bail, P. T., 34, 44, 77, 127, 128
Baker, J. D., 60, 79, 95, 132
Barfield, E. C., 62, 67, 131
Bartell, F. E., 26, 27, 127
Beeson, D. M., 102, 127
Benner, F. C., 26, 27, 127
Bennion, D. W., 63, 67, 127
Bernard, G. G., 15, 27, 130
Bernard, W. J., 100, 127
Blair, P. M., 33, 43, 82, 85, 96, 128
Blum, H. A., 15, 27, 133
Boatman, E. M., 28, 132
Bobek, J. E., 14, 22, 26, 77, 127, 128
Boley, D. W., 96, 130
Bossler, D. P., 28, 43, 130
Botset, H. G., 50-52, 59, 60, 76, 127, 134
Bradley, H. B., 51, 60, 127
Breston, J. N., 44, 133
Bridges, P. M., 43, 77, 127, 134
Brons, F., 96, 127
Brown, H. W., 19, 27, 127
Brown, R. J. S., 27, 127
Brown, W. O., 34, 44, 127
Brownscombe, E. R., 27, 128
Bruce, W. A., 19, 27, 133
Buchwald, R. W., 96, 127
Buckles, G. L., 77, 127
Buckley, S. E., 23, 27, 31-35, 43, 81, 82, 85, 87, 88, 127
Buckwalter, J. F., 77, 127
Bul'chuk, D. D., 77, 128
Burdine, N. T., 24, 26, 127
Burton, M. B., Jr., 51, 52, 59, 127
Bush, J. L., 96, 127

C

Calhoun, J. C., Jr., 12, 17, 26, 27, 33, 44, 78, 79, 81, 84-86, 95, 127, 128, 130, 131, 134
Calloway, F. H., 94, 96, 127
Canadian Gulf Oil Co., 44, 129
Cardwell, W. T., Jr., 43, 127
Carll, J. F., 9, 11, 127
Carpenter, C. W., Jr., 60, 72, 73, 77, 82, 96, 100, 128, 133
Caudle, B. H., 27, 44, 47, 50-52, 54, 56-61, 70, 76, 80, 81, 85, 93, 95, 98, 100, 102, 108, 109, 115, 117, 121, 124, 127-129, 131, 133, 134
Chatenever, A., 27, 128
Cheek, R. E., 51, 52, 60, 128
Christensen, R. J., 102, 130
Chuoke, R. L., 44, 128
Clark, K. K., 68, 128
Cobb, T. R., 95, 129
Cole, F. W., 77, 131
Collins, R. E., 61, 77, 128, 132
Cooper, H. E., Jr., 52, 60, 95, 109, 131
Corey, A. T., 26, 28, 128
Cosgrove, J. J., 77, 82, 85, 96, 134
Cotman, N. T., 61, 128
Craig, F. F., Jr., 27, 45-47, 50-52, 56, 59, 60, 72, 73, 76, 77, 81, 85-88, 91-93, 96, 100, 102, 115, 124, 128, 129, 131-133
Crawford, D. A., 60, 133
Crawford, P. B., 51, 52, 56-61, 67, 77, 127-129, 131-134
Croes, G. A., 44, 77, 128, 129
Crowell, D. C., 44, 128

D

Dalton, R. L., Jr., 100, 128
Davis, G. T., 26, 128
Dean, G. W., 44, 128
Denekas, M. O., 13, 26, 127, 128
Deppe, J. C., 53, 60, 81, 85, 95, 128
Devlikamov, V. V., 72, 77, 128
Dietz, D. N., 34, 44, 128
Dodge, C. F., 67, 130
Doepel, G. W., 95, 128
Donahue, D. A. T., 68, 128
Dougherty, E. L., 60, 128
Douglas, J., Jr., 43, 82, 83, 85, 96, 128
Dunning, H. N., 27, 130
Dyes, A. B., 44, 51-53, 55, 56, 58, 60, 61, 95, 102, 108, 128
Dykstra, H., 28, 47, 64, 65, 68, 79, 81, 82, 84-90, 92, 95, 114, 115, 123, 124, 129

E

Earlougher, R. C., 77, 83, 85, 87, 96, 127, 129
Edgerton, G. H., 77, 127
Efros, D. A., 43, 129
Elkins, L. F., 23, 27, 63, 67, 68, 102, 129
Engelberts, W. L., 34, 44, 129
Erickson, R. A., 51, 52, 60, 61, 95, 108, 128

F

Farouq Ali, S. M., 57, 60, 133
Fatt, I., 24, 27, 28, 127, 129, 133
Fay, C. H., 51, 59, 129
Fayers, F. J., 43, 129, 130
Felsenthal, M., 84-86, 95, 96, 129
Fernandes, R. J., 95, 127
Ferrell, H., 58, 61, 129
Fettke, C. R., 11, 129
Fischer, M. J., 57, 61, 131
Fitch, R. A., 79, 95, 102, 129
Flock, D. L., 51, 60, 98, 100, 132
Forrester, C. E., 102, 133
Fried, A. N., 102, 129
Funk, E. E., 77, 129

G

Gardner, G. H. F., 134
Gatlin, C., 102, 129
Gaucher, D. H., 72, 73, 77, 129
Geertsma, J., 77, 129
Geffen, T. M., 23, 27, 44, 47, 51, 59, 76, 81, 85, 87, 96, 115, 124, 128, 129
Goddin, C. S., Jr., 73, 77, 129
Gogarty, W. B., 102, 129
Gonzalez, H. J., 67, 127
Gottfried, B. S., 43, 129

Gournay, L. S., 27, 133
Greenberger, M. H., 83, 85, 96, 129
Greenkorn, R. A., 62, 67, 129, 130
Griffith, J. D., 79, 95, 102, 129
Griffiths, J. C., 63, 67, 127
Groult, J., 63, 67, 129
Guckert, L. G., 51, 52, 60, 95, 129
Guerrero, E. T., 28, 83, 85, 87, 96, 129, 131
Guilinger, W. H., 43, 129
Gurses, B., 96, 129
Guthrie, R. K., 83, 85, 96, 129

H

Habermann, B., 51, 60, 130
Hafford, J. A., 23, 27, 132, 133
Hall, H. N., 43, 134
Harpke, C. F., 102, 131
Hartsock, J. H., 61, 130
Hassler, G. L., 23, 27, 130
Hauber, W. C., 81, 85, 95, 130
Haynes, G. W., 44, 129
Hazebroek, P., 61, 132
Helander, D. P., 96, 127, 129, 130
Hele-Shaw, H. S., 59, 130
Heller, J. P., 51, 60, 127
Hemstock, R. A., 28, 131
Henderson, J. H., 28, 60, 128, 132
Hendrickson, G. E., 85, 88, 92, 96, 130
Henley, D. H., 60, 96, 130
Heuer, G. J., 95, 129
Hiatt, W. N., 82, 85, 96, 130
Hickman, B. M., 50, 59, 95, 128
Hickman, W. B., 27, 132
Hickok, C. W., 102, 130
Higgins, R. V., 82, 85-88, 90, 93, 96, 130
Hocott, C. R., 77, 130
Holbrook, O. C., 15, 27, 130
Holmgren, C. R., 43, 44, 130
Hough, E. W., 26, 130
Hovanessian, S. A., 43, 130
Hunter, Z. Z., 61, 130
Hurst, W., 51, 59, 80, 85, 95, 130
Hutchinson, C. A., Jr., 60, 63, 67, 73, 77, 130

I

Imperial Oil Ltd., 44, 131
Irby, T. L., 61, 129

J

Jacquard, P., 44, 67, 130, 132
Jahn, J. O., 67, 130
Jain, C., 67, 130
Janzen, H. B., 51, 60, 132
Jenks, L. H., 27, 131
Jennings, H. Y., Jr., 27, 130
Jewett, R. L., 60, 79, 95, 132
Johansen, R. T., 27, 130
Johnson, C. E., Jr., 85, 95, 124, 130
Johnson, C. R., 63, 67, 129, 130
Johnson, E. F., 24, 28, 43, 44, 130, 133
Johnson, J. P., 79, 85, 95, 130
Johnson, N., 27, 130
Jones-Parra, J., 27, 34, 44, 130
Jordan, J. K., 77, 130

K

Kantar, K., 96, 130
Kelley, D. L., 44, 131
Kemp, C. E., 58, 61, 129
Kennedy, E. L., 28, 131
Kereluk, M. J., 77, 131
Kern, L. R., 33, 43, 131
Kerver, J. K., 27, 132, 133

Killins, C. R., 17, 27, 131
Kimbler, O. K., 52, 60, 95, 109, 131
Klinkenberg, L. J., 34, 44, 129
Koeller, R. C., 60, 131
Kruger, W. D., 62, 67, 131
Krutter, H., 52, 60, 131
Kufus, H. B., 85, 96, 131
Kyte, J. R., 27, 28, 42, 44, 131

L

Lamoreaux, W. E., 27, 43, 132
Lancaster, W. R., 96, 134
Land, C. S., 28, 131
Landrum, B. L., 56-58, 60, 61, 128, 131-133
Law, J., 64, 68, 131
Leach, R. O., 26, 28, 102, 131, 134
Leas, W. J., 27, 34, 44, 60, 82, 96, 131, 133
Lee, B. D., 59, 131
Leeman, E. H., 27, 130
Leighton, A. J., 82, 85-88, 90, 93, 96, 130
Lents, M. R., 63, 68, 131
Leverett, M. C., 17, 19, 23, 27-29, 31-35, 43, 81, 82, 85, 87, 88, 96, 125, 127, 131
Levine, J. S., 33, 43, 131
Lewis, W. B., 28, 131
Lindley, D. C., 72, 73, 77, 129
Loncaric, I. G., 51, 60, 95, 98, 100, 128
Loomis, A. G., 44, 128
Lynch, E. J., 85, 96, 131

M

Martin, J. C., 43, 131
Mattax, C. C., 26, 27, 127, 128, 131
Matthews, C. S., 11, 57, 60, 61, 67, 79, 95, 97, 100, 131-133
McCardell, W. M., 77, 130, 131
McCarty, D. G., 62, 67, 131
McEwen, C. R., 43, 131
McGhee, J., 28, 132
McLatchie, A. S., 24, 28, 131
McNiel, J. S., 51, 60, 132
Menzie, D. E., 51, 52, 60, 77, 128, 131
Messer, E. S., 28, 131
Miller, M. G., 63, 68, 131
Montadert, L., 67, 129
Moore, A. D., 59, 131
Moore, D. W., 76, 128
Morel-Seytoux, H. J., 82, 85, 96, 132
Morse, R. A., 27, 43, 44, 47, 51, 59, 81, 85, 87, 96, 115, 124, 128-130, 132-134
Mortada, M., 56, 60, 132
Moss, J. T., 51, 60, 132
Mungan, N., 22, 27, 28, 44, 132
Muskat, M., 12, 26, 45, 47, 50-54, 59, 60, 69, 76, 79, 80, 85, 93, 95, 96, 124, 132, 134

N

Naar, J., 28, 56, 60, 132
Nabor, G. W., 50, 56, 59, 60, 132, 133
Naumann, V. O., 27, 28, 43, 130, 131
Neilson, I. D. R., 51, 60, 98, 100, 132
Nettle, R. L., 28, 132
Newcombe, J., 28, 132
Nielsen, R. F., 27, 131
Nobles, M. A., 51, 60, 132
Nutting, P. G., 13, 26, 132

O

Odeh, A. S., 51, 60, 127

Osoba, J. S., 27, 132, 133
Ortloff, G. D., 102, 127
Overbey, W. K., Jr., 68, 132
Owens, W. W., 24, 27, 28, 33, 43, 102, 129, 132

P

Pan American Petroleum Corp. (Amoco Production Co.), 16, 27, 129
Parrish, D. R., 27, 43, 44, 102, 129, 132
Parsons, R. L., 47, 64, 65, 68, 79, 81, 82, 84-90, 92, 95, 114, 115, 123, 124, 129
Patton, E. C., Jr., 63, 68, 132
Paulsell, B. L., 59, 97, 100, 132
Peaceman, D. W., 82, 96, 128
Perkins, F. M., Jr., 27, 77, 132, 133
Pfister, R. J., 102, 132
Pickell, J. J., 27, 132
Pinson, J., 61, 128, 132, 133
Pirson, S. J., 12, 26, 28, 132
Polasek, T. L., Jr., 67, 130
Pottier, J., 44, 132
Prats, M., 51-54, 56, 57, 59-61, 79, 85, 87-90, 95, 129, 132
Price, H. S., 66, 68, 134
Pruitt, G. T., 61, 129
Purcell, W. R., 26, 27, 133

R

Rachford, H. H., Jr., 44, 82, 96, 128, 133
Rahme, H. D., 65, 68, 79, 84, 85, 95, 133
Ramey, H. J., Jr., 50, 51, 59, 60, 85, 87, 88, 95, 96, 127, 133
Ramsay, H. J., Jr., 102, 130
Rapoport, L. A., 28, 34, 44, 56, 60, 77, 82, 85, 96, 100, 128, 131, 133
Rathjens, C. H., 28, 128
Raza, S. H., 26, 133
Reiss, L. H., 67, 129
Rice, R. R., 27, 130
Richardson, J. G., 22, 23, 27, 44, 77, 132, 133
Riches, W. W., 27, 127
Roberts, T. G., 85, 96, 133
Root, P. J., 77, 133
Rose, W. R., 19, 27, 133
Rosenbaum, M. J. F., 97, 100, 133
Rough, R. L., 68, 132
Russell, C. D., 27, 131
Russell, D. G., 11, 67, 131
Rzasa, M. J., 26, 28, 130, 132

S

Sandberg, C. R., 27, 133
Sanderlin, J. L., 76, 128
Sandiford, B. B., 102, 133
Sandrea, R. J., 57, 60, 133
Saraf, D. N., 28, 133
Sarem, A. M., 28, 133
Schauer, P. E., 83, 85, 96, 133
Scheidegger, A., 44, 133
Schiffman, L., 44, 133
Schmalz, J. P., 65, 68, 79, 84, 85, 95, 133
Schoeppel, R. J., 95, 133
Schrider, L. A., 81, 85, 96, 134
Schwarz, N., 44, 77, 128, 129
Scott, J. O., 102, 133
Sheffield, M., 43, 133
Sheldon, J. W., 43, 60, 128, 129
Sibley, W. P., 95, 128

Silberberg, I. H., 50, 52, 59, 60, 95, 128, 134
Simmons, J., 61, 128, 132, 133
Sippel, R. F., 27, 133
Skiba, F. F., 77, 133
Skov, A. M., 63, 67, 68, 102, 129
Slider, H. C., 85, 87, 90, 95, 133
Slobod, R. L., 15, 27, 47, 51, 52, 59-61, 95, 102, 128-130, 133
Smith, A. E., 96, 127
Snell, R. W., 28, 133
Snyder, R. W., 43, 85, 87, 88, 96, 129, 133
Stallenberger, L. K., 67, 129
Stanclift, R. J., Jr., 44, 131
Stephan, S. C., Jr., 44, 131
Stewart, C. R., 24, 27, 133
Stiles, W. E., 65, 68, 73, 77, 79, 81, 84-92, 95, 127, 134
Still, G. R., 56, 60, 61, 128, 134
Strickler, W. R., 60, 133
Suder, F. E., 78, 79, 84-86, 95, 134
Sukhanov, G. N., 77, 128
Swanson, B. F., 27, 132

T

Taber, J. J., 102, 134
Tek, M. R., 77, 129
Terwilliger, P. L., 27, 32, 43, 132, 134
Testerman, J. D., 63, 66, 67, 134
Torrey, P. D., 77, 134
Tosch, W. C., 102, 129
Treiber, L. E., 26, 133

V

van der Poel, C., 44, 128, 134
van Everdingen, A. F., 96, 127
van Meurs, P., 34, 44, 128, 134
van Wingen, N., 27, 130

W

Wagner, O. R., 26, 28, 102, 131, 134
Wagner, R. J., 43, 82, 85, 96, 128
Warren, J. E., 66, 68, 77, 82, 85, 96, 134
Wasson, J. A., 81, 85, 96, 134
Watson, R. E., 52, 60, 134
Wayhan, D. A., 96, 134
Welge, H. J., 23, 24, 27, 32, 33, 43, 56, 60, 81, 96, 107, 114, 124, 134
White, P. D., 51, 60, 132
Wilkes, J. O., 77, 129
Wilsey, L. E., 43, 134
Witte, M. D., 54, 60, 70, 76, 81, 85, 93, 95, 115, 117, 121, 124, 128
Wood, B. B., Jr., 26, 130
Wood, H. W., 102, 131
Woods, E. G., 67, 130
World Oil, 111, 130
Wyckoff, R. D., 50-52, 59, 60, 71, 76, 132, 134
Wygal, R. J., 28, 60, 132
Wyllie, M. R. J., 28, 128, 134

Y

Young, J. W., 28, 131
Yuster, S. T., 27, 78, 81, 84, 85, 95, 96, 129, 132, 134

Z

Zeito, G. A., 68, 134

Subject Index

A

Adsorption equilibrium, 14
Aerial photography, 63, 68, 128
Alcohol flooding or slug process, 101, 102, 129
Alundum, 42, 43
American Petroleum Institute:
  Project 47B, 15
  Subcommittee on Recovery Efficiency, 83
Analog models, 60
Areal coverage, 45, 46, 53, 54, 56, 58, 70, 97, 99, 101
Areal sweep efficiency, 10, 45-61, 71, 78-81, 95, 98, 99, 116-118, 122, 125, 132, 133
  After water breakthrough, 53, 120, 121
  At water breakthrough, 50-53, 116, 122
  Definition of, 48
  Factors affecting, 56-58
  Measurement of, 48-50
  Prediction methods, 54-56

B

Bandera sandstone, 42
Barriers: rectilinear impermeable, 57
Bartlesville sand, 9
Bentonitic muds, 22
Berea core, 18, 19, 42, 43
Bibliography, 127-134
Blotter-type model, 48, 50, 59, 133
Bodcaw reservoir, 63, 68, 131
Boise sandstone, 42
Bottom-hole injection pressures, 94, 125
Bottom water coning, 68, 129
Bradford field, 9, 11, 129
Bradford sandstone, 13, 22, 127
Bubble point, 44, 94, 125, 128
Buckley-Leverett: displacement theory, 43, 96, 127, 133
  Models, 87, 88
Bypassing, 73, 94, 128

C

California sands, 93, 115
Canada, 9
  Redwater field or pool, 42-44, 129
Capillarimetric method, 14, 27, 130
Capillary desaturation, 27, 131
Capillary effects or forces, 43, 69, 72-75, 78, 129, 130
Capillary equilibrium, 27, 130
Capillary pressure, 17-19, 33, 43, 127, 131
  Air-brine, 19
  Air-mercury, 19, 20, 27, 132
  Characteristics and effects, 9, 12, 24, 27, 32, 82, 96, 128, 129
  Curve, 17-20
  Data and measurement, 26, 44, 127, 133
  Definition, 29, 103
  Gas-oil, 19
  Gas-water, 17
  Gradient, 29, 31, 32, 104
  Oil-air, 27, 132
  Oil-water, 17, 18
  Water-oil, 17, 19, 29
Capillary viscous force ratio, 74
Carbon dioxide, 101, 102, 127
Carbonated waterflood, 102, 130, 133
Cardium sandstone, 66, 67
Casing leaks, 94

Cementation, 62
Cementing material, 12
Channel flow, 15, 16, 96, 129
Chloroform: for rock sample extraction, 15
Circle flooding, 9
COFCAW process, 101, 102, 132
Coefficient: of permeability variation, 64, 65, 71, 79, 86
Colorado, 84
Compaction, 62
Composite injection, 93, 115-123
Composite water-oil ratio, 114-123
Compressibility: of reservoir rock, 28, 131
Computer: digital, 62, 83
  Program, 81, 82, 88, 90, 93, 123
Condition ratio, 98, 99
Conductance ratio, 55, 117, 126
Conductivity: of five-spot network, 60, 133
  Of pattern, 61, 131
  Ratio, 73
Confidence level, 76
Conformance factor, 63, 92
Connate water, 16, 19, 22, 35, 36, 40, 62, 78, 79, 83, 87, 93, 102, 112, 118-120, 131, 133
  Average values, 20, 24
  Definition, 25, 126
  Irreducible, 26
  Mobility of, 34, 35, 37, 38, 44, 46, 127
Contact angle, 12-15, 126
  Apparent, 15
  Cell, 13
  Measurements, 13, 14, 22, 26, 131
Contamination: of core sample, 22
Core analysis, 9, 64, 66, 114
Core lithology, 66, 67
Core samples, 14, 15, 18-22, 27, 62, 79, 132
Correlation coefficient, 83
Correlation factors, 56
Correlations: areal sweep efficiency at breakthrough versus mobility ratio, 81
  Areal sweep increase after water breakthrough, 53, 81
  Areal sweep and water cut, 79
  Effect of mobility ratio on displaceable volumes injected, 108-110
  Effect of mobility ratio on oil production, 108, 109
  For calculating five-spot water injection rates, 115
  For estimating waterflood performance from histories of floods, 84
  Laboratory, for various flooding patterns, 56
  Mobility ratio versus breakthrough areal sweep efficiency, 46
  Of Lorenz coefficient with gas volume filled up at first production increase, 83
  Of resistance within an electrolytic analog of five-spot pattern, 56
  Pore size distribution and permeability, 12
  Relating linear and five-spot recoveries, 82
  Water drive recovery for various lithologies, 83
  Waterflood recovery and both mobility ratio and permeability distribution, 79
Cotton Valley field, 63, 68, 131
Critical saturation, 38
Cross flooding, 56, 60, 77, 134
Crossbedding, 94

Crossflow: between layers, 66, 67, 69, 72-74, 78, 82, 85, 90, 93, 94, 126, 129, 133
  Index, 74
Curtailment: effect on waterflood recovery, 77, 131, 134
Cycling, 59, 63, 68, 129, 131

D

D-3 reef, Redwater field, 42, 43
Damaged well, 97
Darcy's law, 45, 50, 103
Density difference: water-oil, 29, 31, 72, 73, 75, 103, 126
Depositional environments, 62, 63
Detergents, 102, 134
Detrital material, 12
Digital models, 60
Dipping reservoirs, 60, 133
  See also Formation dip
Dispersed feed procedure, 23
Displaceable volume, 52, 54, 55, 108-110, 126
Displacement: liquid-liquid, 44, 128
  Of oil by air, 15
  Of oil by gas, 104
  Of oil by water, 17, 27, 39, 44, 77, 81, 104, 128, 129, 131, 134
  Of water by oil, 15, 27
  Pressure, 17, 19
  Ratio, 14
  Tests, 15
Displacement efficiency, 52, 55, 56, 75, 95
  Of gas, 44, 128, 129
  Of oil, 10, 26, 28, 67, 75, 95, 101, 102, 125, 129, 134
  Of oil by water, 29-44
Displacing phase (fluid) saturation, 23, 107
Dissolved gas drive, 92
Dolomitization, 10, 62
Domes unit, 102, 133
Drainage: capillary pressure curve, 18, 19
  Cycle, 18, 19
  Definition of, 17
  Relative permeability characteristics, 20, 23
Drilling muds, 14

E

Economic limit: of waterflood, 16
  Of water-oil ratio, 94
Edge water: bypassing of, 44, 128
Edwards limestone, 19
Efficiency:
  See specific type
Electric analyzer, 62
Electrical resistivity measurements, 23, 24, 28, 132
Electrolytic model, 48, 50-52, 59, 127, 133
End to end flooding, 58, 93
Enriched gas drive, 63
Equations: areal sweep efficiency, 116, 117
  Average water saturation, 33, 36, 114
  Average water saturation at breakthrough, 35, 37
  Capillary pressure, 17, 103
  Capillary pressure gradient, 29
  Coefficient of variation, 64
  Crossflow index, 74
  Cumulative injected water volume, 33, 36

Darcy's law, 103
Density difference, 103
Displacing phase saturation, 23, 107
Fractional flow, 29, 30, 35, 103, 104
Fractional recovery efficiency, 83
Frontal advance, 32, 105, 106
Gas balance, 40
Geometric mean permeability, 66
Incremental oil produced from newly swept region, 118
Incremental time occurring from interference to fillup, 117
Incremental water produced from swept region, 118
Initial gas saturation, 98, 116, 122
Injected water at breakthrough, 116
Injected water at fillup, 116
Injected water at interference, 116
Injectivities for regular patterns with unit mobility ratio, 55
J-function, 19
Mobility ratio, 45, 114
Oil bank pressure, 40
Oil producing rate, 122
Oil-water ratio, 30
Outer radius of oil bank, 116
Outer radius of waterflood front, 116
Permeabilities in parallel, 66
Permeabilities in series, 66
Permeability variation, 65, 114
Pore volume, 116
Pore volumes of cumulative injected water, 114
Pressure level at which trapped gas is dissolved, 40
Producing water-oil ratio, 79, 118
Rate of mass accumulation of water in rock element, 105
Stock-tank oil in place, 114
Total distance a plane of given water saturation moves, 32, 35, 37
Total oil recovery, 114
Total velocity, 103
Volume balance, 40
Volume of water injected to water breakthrough, 118
Volumetric sweep efficiency, 69
Water drive recovery, 83
Water flow rate, 105
Water injectivity, 70, 116, 117, 122
Water saturation as function of distance and time, 105
Young-Dupre, 12
Equilibrium contact angle, 14
Equipotential distributions, 48
Evaporation method: for determining interstitial water, 28, 131
For determining connate water saturation, 25, 28
External drive techniques, 23, 24
External gas drive, 24, 33

F

Faulting: in reservoirs, 62
Fillup, 53, 59, 69, 75, 79, 81, 83, 93, 98, 116-118, 122
Film: polyethylene or Saran, 22
Five-spot pattern, 9, 45, 82, 85, 93, 112
  Areal sweep efficiency, 46, 80
  Characteristics of, 48-61
  Effects of mobility ratio, 108, 127, 130
  Effects of off-pattern wells, 57, 61, 132
  In dipping reservoirs, 60, 133
  Recovery histories, 73, 131
  Variations in injectivity, 70, 71, 78, 81
  Volume sweep efficiency, 72
  Waterflood performance, 76, 77, 79, 88, 96, 115-124, 129, 134
Flood front, 35, 38, 58, 78, 116
  Advance of, 48, 74
  Flowing oil behind, 79, 80
  Producible oil behind, 56

Radius, 70, 71
Relative permeability to oil ahead of, 114
Saturation gradient behind, 45, 46
Water, 53, 75, 122
Flood pot tests, 22, 23, 79, 92
Flooding patterns, 45, 47-61, 79, 80, 95, 99, 112, 127-129
Floodout, 16, 20, 22, 23, 26, 46, 54, 92, 116, 126
  Patterns, 67, 131
Flow capacity, 56, 60, 63, 64, 76, 79, 81, 98, 132
Flow models, 45, 77, 98, 128, 132, 133
Flow properties: of reservoir rock, basic water-oil, 12-28, 38
Flow tests, 11, 27, 67, 82, 131, 133
Fluid conductivity, 78
Fluid displacement: mechanism in sands, 27, 43, 96, 127
Fluid distribution: in water-oil systems, 19, 23
  Within rock pore spaces, (porous systems), 15-17, 27, 129
Fluid flow models, 50-53, 60
Fluid-fluid interfaces, 60, 128
Fluid injection projects, 61, 64, 132
Fluid injectivity, 53, 56-58, 70, 81
Fluid level surveys, 94
Fluid mappers, 48, 51, 52, 59, 60, 128, 131, 132
Fluid property data, 9
Fluid samples, 10
Flush oil production, 58
Foam-drive process, 102, 129
Foam injection, 101, 102
Formation dip, 39, 57-59, 61, 126, 131
Formation outcrops: information from, 63, 67, 129
Formation volume factor, 40, 125
Formation water, 13
Forest Oil Corp., 9
Foster field, 88, 91
Fosterton field, 68
Four-spot pattern, 48-50, 52-54, 56, 59, 80, 95, 128
Fractional flow curve, 30-32, 35-40, 92, 107, 112, 113
Fractional flow equation, 29-31, 33, 45, 103, 104, 112
Fractional oil recovery, 93, 114
Fractional recovery efficiency, 83
Fractures: elliptical, 61, 132
  Horizontal, 58, 61, 99, 128, 131
  Orientation, 68, 129, 132
  Vertical, 58, 61, 128, 130, 133
Free gas, 43, 44, 53, 60, 92, 94, 100, 116-118, 121-123, 130-132
Frontal advance: equation, 31-33, 81, 105, 106
  Practical applications, 35-38
  Rate, 81
  Theory, 23, 29-34, 81
Fry pool, 9

G

Gamma ray log, 66, 67
Gas cap, 69
Gas condensate cycling, 63, 68, 131
Gas-cut cores, 25
Gas cycling performance, 63
Gas injection, 45, 47, 68, 78, 102, 132
Gas-oil ratio, 23, 40
Gas phase: pressure in, 17, 125
  Trapping of, 42
Gas saturation: before waterflooding, 25, 94, 112
  Initial, 39-43, 53, 54, 71, 78, 80-82, 85, 97-99, 116, 122, 126
  Trapped, 39-43, 94, 116, 122, 126
Gas solubility, 40
Gas-water injection, 101, 102, 128
Gelatin model, 48, 51, 52, 59, 127
Geological engineering, 10
Geological zonation, 66

Glass-bead pack, 17, 56, 82
Gravity drainage, 32, 43, 97, 134
Gravity forces, 43, 71, 72, 75, 78, 104, 129-131

H

Hafford method, 23
Hassler method, 23
Hele-Shaw models, 48
Hysteresis, 17, 18, 20

I

Illinois Basin waterfloods, 83
Imbibition: capillary pressure curve, 18, 19
  Definition of, 17
  Displacement tests, 14, 22
  From a glass capillary, 15
  Oil, 14
  Relative permeability characteristics, 20, 28, 131
  Water, 14, 75
Immiscible displacement, 96, 128
Immiscible fluids, 17, 44, 48, 132
In-situ combustion, 59, 60, 132, 133
Inertial effects, 23
Injected-fluid coverage, 96, 130
Injected fluid front, 69
Injection rate, 34, 93, 100, 101, 121, 122
  At water breakthrough, 54
  Effect of mobility ratio, areal sweep and pattern, 60, 95, 128
  Predictions of, 95, 132
  Variations, 81, 117
  Versus time, 90, 91
  Volumetric, 79
Injection surveys, 94
Injectivity: correlation, 93
  Ratio, 70, 71
  To oil, 121
Interface pressure, 125
Interfacial energy, 12
Interfacial tension: air-oil, 15
  At reservoir pressure and temperature, 26, 130
  In waterflooding, 28, 126, 131, 132
  Oil-water, 12, 17, 28
  Water-oil, 15, 24, 102
Interference: definition of, 116
Intermittent gas drive, 23
Internal gas injection, 68, 132
Interstitial heterogeneity, 68, 131
Interstitial water saturation, 25, 28, 131
Invasion efficiency, 69, 125
Inverted well patterns, 48, 49
Irreducible connate water saturation, 25-26
Isobar maps, 9
Isopach maps, 9
Isopotential distribution, 79
Isopropyl alcohol, 102

J

J-function, 19
Jourdanton field, 19

K

Kozeny-Carmen equation, 134

L

Laboratory measurements: of fluid samples, 10, 22
Laboratory tests: involving imbibition and displacement procedures, 14
  Models, 81, 100, 128
  Of relative permeability, 25
Laplace equation, 50
Layered reservoir concept, 63
Layer permeability ratio, 74
Layer selection, 75

Length: of stabilized zone, 34
Line drive, 9, 52, 53, 56, 61, 80, 129
   Direct, 48, 50, 53-58, 81, 82, 108
   Staggered, 48, 53-55, 60, 82, 109, 132
Linear displacement equation, 43, 131
Linear flow tests, 79
Linear frontal displacement, 96, 131
Linear waterflood behavior, 28, 44, 82
Liquid-filled: injectivity, 121
   Pattern, 121, 122
Liquid-liquid displacements, 44, 128, 132
Liquid-saturated reservoir, 98
Log normal permeability distribution, 64-66, 76, 82, 84, 87
Lorenz coefficient, 65, 66, 79, 83, 84, 86

M

Macroscopic vertical permeability, 67
Mass transfer: between phases, 32
Mathematical models, 62, 82-84, 90, 93
Mean square difference, 76
Measurements: in laboratory of fluid samples, 10
   Of areal sweep efficiency, 48
   Of capillary pressures, 26, 133
   Of contact angle, 13, 14, 22, 26, 131
   Of electrical resistivity, 23, 24, 28, 132
   Of relative permeability, 27, 28, 129, 133
   Of water-wetting preference, 14
Mercury injection technique, 19, 24
Methanol, 15
Methylene blue, 15
Micellar solutions, 102, 129
Micellar slug flooding, 101
Micro emulsions, 102
Microscopic studies, 15
Miscible displacement, 60, 102, 128, 130
Miscible floods, 50, 74, 79, 95, 102, 129
Miscible fluids, 45, 46, 48, 50, 52, 72, 80
Miscible slug process, 95, 102, 127
Mobility: of connate water, 34, 35, 44
   Of displacing fluid, 46
   Of oil, 38, 45, 79, 80, 126
   Of water, 38, 45, 46, 79, 80, 102, 126
Mobility ratio, 10, 51, 54, 56, 58-60, 80, 88, 93-95, 114, 121, 127
   As function of sweepout pattern efficiency: and displaceable volumes injected, 110
      At various well-producing cuts, 109-111
   Definition of, 45, 125
   Development of concept, 45, 46
   Effect on: displaceable volumes injected, 108-110
      Efficiency of miscible displacement, 60, 130
      Oil recovery performance, 76, 79, 95, 130
      Ratio of producing rate to injection rate, 97, 98
      Reservoir flow patterns, 60, 95, 127
      Volumetric sweep efficiency, 69-72
      Oil production, 60, 70, 95, 108, 109, 127, 128, 131
   Favorable, 46, 53, 59, 70, 71, 73, 74, 82, 86
   Fluid mapper studies, 60, 128
   Of unity, 48, 53-57, 69-74, 76, 81, 82, 85-87, 96, 98, 117, 122, 132
   Other than unity, 50-52, 58, 69-74, 76, 85-87, 98, 114, 115, 117, 122
   Ranges during waterflooding, 46-47
   Resistance network study, 60, 132
   Unfavorable, 46, 53, 59, 70, 71, 73, 74
   Versus area swept, 54, 57, 78
      Areal sweep efficiency, 52-55, 61, 71, 78, 99, 122, 130
      Conductance ratio, 55, 117
   Water-oil, 46, 54, 59, 69, 79, 81, 86, 92, 114

Models: See specific type
Monograph: base permeabilities used, 20
   Objectives of, 10
   Organization of, 10, 11
Multiple regression analysis, 42

N

Native-state cores, 21, 22
Nebraska, 84
Nellie Bly sandstone, 42, 43
Network models, 28, 129
New York: Bradford oil field, 11, 129
Nine-spot pattern, 48, 50, 52-56, 59-61, 80, 95, 109-111, 128, 129, 131, 132, 134
Nonstabilized zone, 32
Non-wetting phase, 14-17, 20
Normal well patterns, 48, 49, 61
   Sweep beyond, 51, 52, 57, 58, 128
North Burbank unit waterflood, 59, 61, 130
Nuclear magnetic relaxation, 15, 27, 127
Nuclear magnetic resonance, 28, 133
Numerical model, 51, 52
   See also mathematical models

O

Off-pattern wells, 57, 61, 132
Oil and Gas Conservation Board of Alberta, 44, 129, 131
Oil bank, 23, 39, 40, 46, 53, 56, 75, 79, 80, 94, 95, 97, 116
   Pressure, 40, 45
Oil-base muds, 22
Oil-carrying channels, 15
Oil cut, 33
Oil displacement: by gas, 104
   By water, 17, 27, 39, 44, 77, 81, 104, 128, 129, 131, 134
   Efficiency, 10, 26, 28, 67, 75, 95, 101, 102, 125, 129, 134
Oil-emulsion muds, 22
Oil-filtrate mud, 25
Oil migration, 97-99
Oil phase: pressure in, 17, 33, 103, 125
Oil productivity, 58, 59, 69
Oil recovery: at water breakthrough, 35, 57, 72
   By crossflooding, 56, 60, 74, 134
   By gas or water drive, 27, 43, 83, 124, 134
   Calculating, 10, 43, 93, 134
   Comparison of: actual and predicted, 89, 90
      Patterns, 61, 128
   Cumulative, 10, 87
   Due to resaturation effects, 94
   Effect of: free gas, 43, 130
      Initial gas saturation, 41
      Permeability variation and mobility ratio, 76, 128
      Rate, 77, 130, 133
      Rock wettability, 38
      Trapped gas, 42, 43, 122
      Viscosity, 96, 129
   Efficiency, 73
   Factors affecting, 94, 95, 101, 122, 123
   In five-spot pattern, 60, 73, 95, 99, 100, 128
   Layer by layer method, 78
   Of gas or water injection, 47, 59, 96, 115, 124, 128, 134
   Performance prediction of 9, 22, 66, 68, 95, 97, 124, 129
   Total, 33, 55, 58, 86, 90, 92, 98, 119-121
   Use of fluid-fluid interfaces to predict behavior, 60, 128
   Versus WOR, 38, 99, 114, 115, 123
Oil saturation, 15, 40, 81, 92, 97, 99, 122
   Initial, 121, 126

Reduction by trapped gas, 42, 43
   Versus capillary pressure 18
Oil shrinkage, 92
Oil-solid interface, 14
Oil-water interface, 34, 50
Oil-wet: definition of, 13
   Relative permeability curves, 21, 46
   Rock, 16-22, 24, 25, 30, 31, 34, 36, 39-42, 75, 95, 101
   Sand grains, 15, 17
   Systems, 14, 31, 35, 38, 46, 47, 102
Oklahoma: Bartlesville sand, 9
   Denver basin, 84
Overburden pressure, 24, 28, 129
Oxygen penetration: packaging of cores to prevent, 22

P

Panhandle field, 88
Parallel-layer flow, 63
Partially depleted reservoirs, 96, 130
Pattern area, 52, 53, 55
Pattern areal sweep, 69, 125, 126
Pattern geometry, 82
Pattern pore volume, 52, 55, 69
Pembina field, Alberta, 66, 67
Penn State method, 23
Pennsylvania: Bradford field, 9, 11, 129
   Bradford sandstone, 13, 22, 127
   Geology of some oil regions, 11, 127
   Pithole City area, 9
Peripheral floods, 58, 78, 82, 93, 96, 129, 130
Permeabilities: in parallel and series, 66
Permeability: absolute, 12, 20, 75, 125
   Anisotropic, 57, 60, 132
   Areal variations, 62, 63, 66, 67
   Averaging, 66
   Coefficient of variation, 64, 65, 71, 79, 86
   Core analysis, 64, 66
   Directional, 56-60, 62, 63, 67, 94, 99, 129, 131
   Effective, 12, 20, 29, 45, 125
   Horizontal, 57, 67
   In-place measurement, 10
   Lateral variation, 63, 93
   Log normal distribution, 64-66, 76, 82, 84, 87
   Of rock or formation, 12, 29, 31, 75, 99, 101
      Average, 112, 125
      Complex variation of, 34
      Contrast between layers, 70, 71, 79
      Development of, 62
      Log normal distribution, 64
      Peripheral-type flood, 58, 59
Permeability: ordering, 65
   Profile, 47, 68, 95, 129, 133
   Relative, 9, 12, 17, 19-21, 24, 28, 33, 81, 83, 92, 127, 133
      Averaging characteristics, 24
      Calculated from displacement experiments, 28, 43, 130
      Calculated from pore size distribution data, 26, 127
      Curves, 20, 21, 25, 26
      Drainage characteristics, 23
      Gas drive method for determining, 43, 132
      Gas-oil, 23, 26, 32, 33, 128
      Gradient, 82
      Imbibition, 28, 60, 132
      Laboratory measured, 15, 22, 25, 27, 128, 129, 132, 133
      Oil-water, 27, 133
      Ratio, 23-25
      Three-phase, 24, 25, 28, 128, 132, 133
      To gas, 27, 125, 128
      To oil, 23, 26, 29, 30, 46, 104, 113, 114, 125, 128

To water, 23, 30, 54, 104, 113, 114, 121, 125
Water-oil, 20-23, 25, 26, 33-35, 45, 47, 75, 78, 82, 91, 92, 112, 115
Specific, 24, 63
Stratification, 63, 64, 66, 72, 95, 96, 130, 132
To water, 46
Trends, 63
Uniform, 56, 66, 72, 74-76
Variation, 64-66, 71, 75, 76, 79, 80, 86, 92, 93, 114, 115, 126, 128
Vertical, 10, 62, 63, 66, 67, 75, 78
Photomicrographs, 15
Pilot floods: advantages and limitations, 97
Pilot model tests, 99
Pilot performance, 11, 59, 96, 99, 127
Pilot waterflooding, 60, 95, 97-100, 128, 133
Pithole City area, 9
Plugged perforations, 94
Polar compounds, 13
Polar impurities, 26, 127
Polymer, 101, 102, 133
Pore geometry, 20, 23, 95
Pore size distribution, 12, 17, 19, 24, 26, 62, 78, 127
Porosity, 10, 12, 24, 63, 83, 87, 100, 105, 112, 122, 126
Development of, 13, 56, 62
Distribution, 95, 130
In-place measurement of, 10
Positional approach, 63, 64
Potentiometric model, 45, 48, 50-53, 59, 60, 97, 98, 131
Pressure: buildup, 11, 67, 98, 131
Differential, 55, 121
Distribution, 33, 45, 48, 82
Drawdown, 98
Gradients, 23, 45, 48, 75, 99
Interference, 68, 129
Pulsing, 101, 102, 132
Transients, 9, 63, 68, 128
Primary phase, 33
Probability permeability distribution, 84
Production engineering: and reservoir engineering aspects of waterflooding, 9
Production engineers: responsibilities of, 9
Production logging techniques, 63, 67, 129
Productivity index, 94, 98
Proration, 94
Pseudointerface, 82
Pulse testing, 63, 67, 130

R

Rate sensitivity, 75
Ratio: displacement by oil, 14
Displacement by water, 14
Recoverable hydrocarbon pore volume, 74
Recoverable oil, 87, 88, 98, 99
Recovery factor, 83, 96, 133
Recrystallization: developing porosity by, 62
Redwater cores, 131
Redwater field, 42-44, 129
References: for areal sweep efficiency, 59-61
For basic water-oil flow properties of reservoir rock, 26-28
For conclusion, reservoir engineering aspects of waterflooding, 102
For efficiency of oil displacement by water, 43, 44
For example waterflood calculations, 124
For introduction, 11
For methods of predicting waterflood performance, 95, 96
For mobility ratio concept, 47
For pilot waterflooding, 100

For reservoir heterogeneity, 67, 68
For vertical and volumetric sweep efficiencies, 76, 77
Refined oil: evaluating reservoir wettability with, 14
Regression analysis technique, 62
Relative permeability: See permeability: relative
Relative injectivity, 70
Relaxation rate, 15
Remedial work, 10
Replacement: developing porosity by, 62
Resaturation: effects, 94
Method, 86
Reservoir engineering: and production engineering aspects of waterflooding, 9
Early stages of, 13
Elements and fundamentals, 26, 127, 129, 132
Studies to permit better evaluation of pilot floods, 97
Usage of permeability variation equation, 65
Reservoir engineers: confidence in frontal drive equation, 33
Philosophy of insulated layers, 67
Responsibilities of, 9, 10
Reservoir heterogeneity, 10, 24, 34, 59, 62-69, 78, 97, 99, 101, 129, 134
Reservoir-scale fractures, 62, 63, 101
Reservoir volume factor: oil, 40, 83, 91, 92, 112, 114
Residual oil saturation, 14, 22-24, 34, 39, 41-43, 53, 81, 87, 92-94, 126
Resistance network, 60, 132
Resistance-type models, 50, 51, 53
Restored-state cores, 21, 22
Rock flow model, 51
Rules of thumb: for predicting waterflood performance, 83
Wettability preferences, 20
Rushford sandstone, 42

S

San Andres pilot waterflood, 96, 130
Saturation distribution: during waterflooding, 32, 33, 37, 38
Saturation gradients, 13, 24, 29, 31, 37, 42, 45, 54, 81, 85, 107
Saturation pressure, 95, 112, 125
Saturation profile, 39, 40
Scaled models, 72, 77, 129
Scaling factor, 19, 77
Scaling laws, 77, 132, 133
Seven-spot pattern, 48-53, 55, 60, 80, 82, 95, 129
Shale breaks or streaks, 67, 68, 134
Shape factor, 82, 96, 129, 130
Short circuits, 69
Single-core dynamic, 23
Solution: changing reservoir characteristics, 62
Solution gas drive, 39, 53, 56, 69, 78, 83, 86, 94, 114
Solvent extraction, 22
Solvent slug, 102
Spatial variations, 85
SPE symbols, 69
Spraberry Trend, 63, 101, 102, 129
Stabilized zone, 32-34, 50, 112, 118, 130
Standard letter symbols for reservoir engineering, 45
Stationary liquid procedure, 23
Statistical reservoir zonation technique, 66, 67, 134
Steady-state tests, 23, 24
Steam distillation, 102
Stiles method, 65
Stochastic model, 67, 127
Stock-tank oil, 94, 114, 116, 119
Storage capacity, 63
Straight-line permeability distribution, 84
Stratification: of vertical permeability, 63

Stratified model, 72
Stratified reservoir, 85, 86, 95, 96, 102, 115, 127-129, 131, 133, 134
Stream tubes, 81, 82
Streamline-channel flow method, 82
Streamline distributions, 48, 50, 79
Structure maps, 9
Subordinate phase, 33
Surface area, 12
Fractional oil-wet, 15
Surface equilibrium, 26, 131
Surface joint patterns, 68, 132
Surface resistance, 59, 130
Surfactants, 13, 14, 22, 101, 102
Sweep efficiency: areal, 10, 45, 47-61
Vertical, 10, 45, 69-77, 82, 97, 116, 125
Volumetric, 45, 69-77, 94, 95, 101, 125
Sweepout pattern efficiency, 54, 57, 109-111
Symmetrical well patterns, 48

T

Tensleep sandstone, 13, 18, 19
Texas: carbonate reservoirs, 88
Fry pool, 9
Jourdanton field, 19
Spraberry Trend, 63, 101, 102, 129
Woodbine cores, 27, 133
Textural heterogeneity, 67
Three-spot pattern, 49, 50
Threshold pressure, 17, 19
Torpedo sandstone, 42, 43
Transparent three-dimensional models, 44, 134
Triple-valued saturation, 31, 32, 43, 127
Two-spot pattern, 49, 50

U

U. of Oklahoma, 15
Ultimate waterflood recovery, 91, 92
United States, 9
Unsteady state method: of measuring three-phase relative permeability, 28, 133
Unswept area, 71

V

Venango sandstone, 17, 18, 42
Vertical coverage, 46, 73, 74, 82, 101
Vertical sweep efficiency, 10, 45, 69-77, 82, 97, 116, 125
Definition of, 69
Viscosity: of crude, 27, 130
Of fluid, 23, 33, 45, 82, 104, 129
Of oil, 29-31, 35, 38, 39, 42, 45-47, 79, 83, 91, 92, 94-96, 102, 112, 126
Of salt water, 111, 130
Of water, 29, 30, 35, 38, 46, 47, 79, 83, 91, 92, 95, 102, 112, 126
Viscosity ratio: oil-water, 30, 34, 50, 56, 82
Viscous fingering, 34, 35, 44, 134
Viscous effects or forces, 72-75, 78, 82
Volumetric coverage, 99, 102
Volumetric sweep efficiency, 45, 69-77, 94, 95, 101, 125
Volumetric sweep efficiency: crossflow between layers, 73-74
Definition of, 69
Effect of: capillary forces, 72, 73
Gravity forces, 71, 72
Layer selection on, 75, 76
Mobility ratio, 69-71
Rate on, 75
Volumetric water invasion efficiency, 68, 129

W

Water-base muds, 22, 25, 92
Water breakthrough, 45, 48, 74, 81, 94, 99

Areal sweep efficiency at, 78, 116, 120-122
Average water saturation at, 38, 46
Early, 59, 75
Effect of directional permeability on areal sweep efficiency at, 57, 59
Increase in areal coverage after, 97, 122
Performance after, 36, 37, 50-56, 58, 78-80, 97, 98, 118, 119
Performance from fillup to, 116-118
Recovery at, 33-36, 76
Trapped gas saturation versus oil saturation, 43
Water channels, 15
Water cut, 33, 36, 37, 55-58, 84, 88, 92, 93, 114
Water-drive recovery efficiency, 83
Water-drive reservoir, 44, 128
Waterflood oil recovery: factors affecting performance, 94, 95
Waterflood pattern: factors affecting selection of, 58, 59
Waterflood performance: after water breakthrough, 118, 119
From fillup to breakthrough, 117
From interference to fillup, 117
Prior to interference, 116, 117
Waterflood performance predictions:
API statistical study, 83
Aronofsky method, 80, 81
Band method, 81
Buckley-Leverett method, 81
Caudle et al. method, 80
Comparisons of: Craig et al. and Higgins-Leighton methods, 86, 87
Dykstra-Parsons, Stiles, Suder-Calhoun and Felsenthal et al. methods, 84-86
Stiles, Dykstra-Parsons, and Snyder-Ramey methods, 87, 88
Stiles and Yuster-Suder-Calhoun methods, 84
Craig-Geffen-Morse method, 81
Deppe-Hauber method, 81
Douglas-Blair-Wagner method, 82
Douglas-Peaceman-Rachford method, 82
Dykstra-Parsons method, 79
Guerrero-Earlougher method, 83
Guthrie-Greenberger method, 83
Hiatt method, 82
Higgins and Leighton method, 82
Hurst method, 80
Morel-Seytoux method, 82
Muskat method, 79
Prats-Matthews-Jewett-Baker method, 79
Rapoport-Carpenter-Leas method, 82
Schauer method, 83
Stiles method, 79
Warren and Cosgrove method, 82
Yuster-Suder-Calhoun method, 78
Waterflood prediction methods: comparison with actual, 87, 90
Empirical, 83, 84
Practical use of, 93, 94
Primarily concerned with areal sweep, 79-81
Primarily concerned with reservoir heterogeneity, 78, 79

Primarily dealing with displacement mechanism, 81-83
Recommended procedures, 90-93
Waterflood recovery: calculation of ultimate, 114
Waterflooding: development of, 9, 10
First in Texas, 9
First occurrence of, 9
General history of, 9, 10
Relation between reservoir engineering and production engineering, 9
Water injection, 16, 78, 102
Accidental occurrence of, 9
Calculated: for start up to interference, 117
Total, 123
Existence of initial gas saturation prior to, 80
In pressure pulsing operation, 101
Into liquid filled two-layer formation, 70
Predictions for five-spot, 116
Pressure, 63
Pressure transient testing of fractured wells, 68, 128
Oil recovery performance from, 47, 59, 74, 75, 96, 97, 99, 124, 128
Simulating, 53
With free gas initially present, 121, 122
With no free gas initially present, 118-121
Water injection rates, 58, 121, 122
Calculations for five-spot, 115, 123
Effect of tenfold change, 72
Effect of gravity segregation, 75
In pattern flooding calculations, 53
Prediction of, 10, 88
Ratio of, to oil producing rate, 83
To interference, 116, 117
Water injectivity, 54, 56, 59, 69, 79, 93, 97, 99
Water injectivity index, 69
Water invasion, 69
Water-oil contact, 19, 32
Water-oil flow properties, 12-28, 93, 101
Water-oil ratios, 33, 76, 82, 86, 91, 121, 122
Calculation of, 55, 56, 118-120, 124
Composite producing, 90, 92, 93, 114, 115, 123
Early increase from off-pattern wells, 57
Economic limit, 94
Effect of gravity segregation, 75
Performance, 59, 81, 82
Prediction of, 10, 23
Reduction due to trapped gas, 43
Versus average water saturation, 37
Versus fraction of recoverable oil, 87, 88
Versus oil recovery, 38, 74, 79, 99
Water phase: pressure in, 17, 33, 103
Water saturation, 20, 29, 30-33, 35, 78, 94, 105-107, 118, 126
At end of imbibition, 18, 19
At floodout, 79
Average, 36, 37, 46, 81, 114, 119-121, 126

In formation adjoining the hole, 25
Initial, 91, 92, 112
Versus capillary pressure, 18
Versus fractional flow of water, 30, 31, 36, 38-40, 113
Versus relative permeability, 26
Water supply sources, 9
Water treating equipment, 9
Water tonguing, 34
Water wet, 14, 17, 19, 20, 22, 34, 38, 39, 46, 72, 102
Definition of, 13
Porous media or sand, 15, 24, 28, 44, 131, 133
Reservoirs, 35, 95
Rock, 16-22, 24-26, 30, 31, 33, 34, 39-43, 75, 95, 101
Systems, 31, 38, 47, 73
Welch field, 88, 92, 96, 130
Welge equation, 33, 36, 81, 107
Wellbore plugging, 73, 94
Well completion techniques, 62
Well log: responses, 66
Studies, 9
Well logging, 10, 25
Well performance, 90, 93, 123, 124
Well pressure: interference tests, 62
Response data, 67, 130
Transient tests, 10
Well productivity, 98
Well spacing, 57, 58, 83, 112
Well workovers, 10
Wettability, 19, 20, 23, 24, 38, 101
Adjustment in waterflooding, 26, 102, 134
By dye adsorption, 27, 130
By nuclear magnetic relaxation method, 27, 127
Effects in laboratory waterfloods, 27, 28, 132
Formation or porous media, 22, 82
Handling of core samples, 21, 22
Intermediate, 13, 18-20
Methods for determining, 14, 15, 27, 133
Number, 15
Of oil-water-solid system, 13
Of rock, 12-15, 17, 22, 26, 27, 38, 127, 128, 133
Of unconsolidated material, 14, 28, 132
Preferential, 14, 17, 20, 42
Relative-water, 15
Reservoir, 14, 34, 62, 78
Wetting: fluid, 15
Phase: 12, 16-18, 20, 75
Wilcox sandstone, 42, 43
Woodbine cores, 27, 133
Wood's metal, 16
Wyoming: Tensleep sandstone, 13

X

X-Ray shadowgraph, 47, 48, 50-52, 59, 80, 95, 133

Y

Young-Dupre equation, 12